Franklin Martin

Lectures on the treatment of fibroid tumors of the uterus

Medical, electrical and surgical

Franklin Martin

Lectures on the treatment of fibroid tumors of the uterus
Medical, electrical and surgical

ISBN/EAN: 9783337142643

Printed in Europe, USA, Canada, Australia, Japan

Cover: Foto ©berggeist007 / pixelio.de

More available books at **www.hansebooks.com**

LECTURES

ON THE

TREATMENT OF

FIBROID TUMORS OF THE UTERUS

MEDICAL, ELECTRICAL AND SURGICAL.

BY

FRANKLIN H. MARTIN, M.D.

PROFESSOR OF GYNECOLOGY POST-GRADUATE MEDICAL SCHOOL OF CHICAGO; SURGEON
WOMAN'S HOSPITAL OF CHICAGO; GYNECOLOGIST CHICAGO CHARITY HOSPITAL
AND THE POST-GRADUATE HOSPITAL; CHAIRMAN SECTION OF OBSTET-
RICS AND DISEASES OF WOMEN OF THE AMERICAN MEDICAL
ASSOCIATION (1895); PRESIDENT CHICAGO GYNECOLO-
GICAL SOCIETY (1895); ETC., ETC.

AUTHOR OF ELECTRICITY IN OBSTETRICS AND GYNECOLOGY.

CHICAGO
THE W. T. KEENER CO
1897

PREFACE.

It was the lot of the writer of these lectures to have entered the practice of gynecology at the dawn of antiseptic surgery. With the possibilities revealed by this great advance, for fifteen years, the surgeons of civilization have been struggling toward perfection. In a little more than a decade wonders have been accomplished and surgery developed to a science. During this period of unprecedented activity the treatment of fibroids of the uterus has received a large share of attention and may now be said to have reached as high state of perfection as any other branch of surgery. The benign nature of these tumors, their slow development, their small mortality without treatment made patients reluctant to accept the radical operations. Hence minor medical and surgical means were sought and persistently cultivated to the limits of their possibilities. Ergot, electricity, the Battey-Tait operation and ligation of the broad ligaments are minor treatments which have relieved much suffering, cured not a few patients, and saved many from the more dangerous procedure of hysterectomy. In the development of these minor means, not only, but in the cultivation and improvement of the major surgical technique the writer has devoted a greater portion of his professional time for the last ten years. The object of these lectures is to place in permanent shape the outcome of this decade of work and to place in the hands of his friends and students a mirror, as it appears to him, of the present status of the treatment of fibroids of the uterus.

FRANKLIN H. MARTIN, M.D.

September 1, 1896.

LECTURE I.

The designation, " Fibroid Tumors of the Uterus, "
which is employed throughout this work, is used for
convenience, euphony, and because it has become the
established and accepted term in America, for all
myomatous or fibromatous enlargements of the uterus.
These enlargements are also called: Myoma or
Fibromyoma Uteri: Fibrous Tumors; Tumeur Fi-
breuse; Myofibroma Uteri: and Hysteroma and Muscular Tumors of the Uterus.

GROSS ANATOMY.

Fibroid tumors of the uterus vary in size from a
scarcely perceptible enlargement of the uterus, to a
tumor which may weigh more than one hundred
pounds. No definite limit of maximum dimension
can be assigned; some grow rapidly to an enormous
size, while others, under apparently the same conditions, increase slowly and never attain such proportions as to produce deformity.

The gross appearance of fibroids of the uterus
differs as widely as does their size. There is no
accounting for the great variety of shapes assumed
by these benign growths. Each one is a law unto
itself. For convenience we divide them according to
their method of development, which influences their
contour, into:

1. Interstitial.
2. Intramural.

3. Subperitoneal.
4. Submucous.

An Interstitial Fibroid is one in which the new growth is uniformly distributed throughout the body of the uterus, without any large, distinct nuclei of development. The external appearance of such a fibroid is that of a symmetrically enlarged uterus, without external nodular enlargements, or without irregular projections of a submucous nature into the cavity of that organ. These symmetrical tumors often grow to weigh many pounds, and frequently enlarge the cavity of the uterus until it will measure many inches in depth, proportionately increasing the area of the endometrium. I have examined such cases where the uterine canal measured fourteen inches in depth.

Intramural Fibroids are those in which the new tissue is confined to the walls proper of the uterus, but in which several distinct centers of development are apparent to the naked eye. While each center may possess several nuclei, the manner of growth is not uniform throughout the uterine walls. This variety makes the uterus irregular and the direction of the canal uncertain. The separate distinct centers of development, as felt through the uterine tissue, are much firmer and less elastic than the new tissue, which makes the typical interstitial variety. Its cut surface frequently exhibits to the naked eye, white cartilaginous centers surrounded by loose connective tissue; these centers are often easily enucleated from the muscular tissue of the walls of the uterus. From the standpoint of gross anatomy this variety is distinct and unique. The intramural fibroid is much less liable to develop into very large tumors, because of their tendency to become pedunculated.

A subperitoneal Fibroid is one which grows from one or more centers of development, which may become pedunculated, and projects from the walls of the uterus into the peritoneal cavity. The gross appearance of these projections when cut through is that of the intramural variety.

A Submucous Fibroid is one which grows from one or more centers of development and projects from the walls of the uterus into its cavity. This variety resembles the intramural in its gross appearance. It may become pedunculated and then becomes an uterine polypus.

I make the unusual distinction between interstitial and intramural fibroids, not only on account of anatomic reasons, but from a therapeutic and surgical standpoint, since their action under the influence of ergot, electricity and the knife is strikingly different.

In the examination of an unusual number of fibroids, I can approximate the relative percentage of the occurrence of the several varieties as follows: Interstitial 55 per cent., subperitoneal 20 per cent., intramural 15 per cent., submucous 10 per cent.

HISTOLOGIC CHANGES.

Fibromata.—Fibroid tumors of the uterus are rarely fibromata; they are less rarely a mixture of fibromata and myomata; while in a large majority of cases they are pure myomata. According to Bland Sutton, "typical fibromata are generally dense tumors consisting of wavy bundles of fibrous tissue. The bundles are composed of long slender fusiform cells closely packed together. The tissue of the tumors, often arranged in whorls, is permeated by blood vessels." This same writer says: "The difficulty of distinguishing between a myoma, a slowly-growing spindle-celled sarcoma and a pure fibroma is well known to skilled histologists."

Myomata are tumors composed of unstriped muscle fibers. Myomata are composed of long fusiform cells with a rod-like nucleus; the size of the cells vary greatly in different tumors. The bundles of muscle fibers are often interwoven in such a manner that the cut surface presents a characteristic whorled appearance. (Bland Sutton.) The well-developed tumor consists of unstriped muscle fibers mixed with more or less fibrous connective tissue and fusiform cells. (Garri-

gues.) According to Pozzi, on microscopic section
fibromata (fibro-myomata) present smooth muscular
fibers and connective tissue in varying proportions.

Blood Vessels.—The vascularity of fibroid tumors
varies greatly. Hard fibroids composed of an unusual
amount of fibrous material constituted in the main of
long fusiform cells, closely packed together, contain
smaller and fewer vessels than the soft or myomatous
tumors, composed of unstriped muscular tissue-fibers
more loosely packed. The latter are frequently very
vascular, the blood vessels being easily traced into
their interior. The former are often surrounded with
a loose fibrous capsule which frequently contains a
free distribution of blood vessels, from which the
tumor proper draws its nourishment. The moderately
vascular are slow in growth while the soft vascular
tumors develop rapidly. "The vessels that traverse
these soft tumors are often of larger size." says Sutton.
"especially the veins, and furnish a loud systolic bruit
on auscultation."

"Some of these myomata are so richly furnished
with blood vessels that on transverse section they look
not unlike erectile tumors. Indeed Virchow speaks
of them as '*cavernous or telangicctatic myomata.*'
The vessels seen on the cut surface are for the most
part veins. Many excellent examples of the extreme
vascularity of such tumors may be found, and it may
easily be conceived that under varying conditions,
such tumors would alter in size, and in some cases
this has been so marked that the tumor seemed to be
erectile." (Bland Sutton.)

Winkle says: "Ordinarily these tumors are not
very vascular but in exceptional cases, not only the
adjacent tissues, but the tumor itself contains a great
number of large vessels."

Nerves.—According to good authority nerves have
been traced into the substance of fibro-myomata.
According to Winkle, Astruc asserted that he found
them in the parenchyma of a polypus, and Bidder, in
a large fibroid, once found a nerve fiber Hartz

described them and their method of termination in
the nuclei of the smooth muscular fibers. Dupuy-
tren also traced nerve fibers into these growths.

Lymphatics.—Fibroids of the uterus contain lymphatics.

Conclusions.—In the proportion that the histology
of a fibroid tumor approaches in microscopic structure fibromata, does it increase in density, become less
vascular, has a more definite capsule, is less intimately
connected with the muscular tissue of the uterus, is of
much slower growth and is much more liable to be of
the intramural variety.

In the proportion that the histology of a fibroid
tumor approaches in microscopic structure, myomata,
does it decrease in density, become softer and more
vascular, is its capsule more indefinite because the
new growth is more intimately distributed with the
the muscular structure of the uterus, is of more rapid
growth and is much more liable to be of the interstitial
variety.

DEGENERATIVE CHANGES.

Fibrocysts.—Mucoid Degeneration.—According to
Bland Sutton large uterine myomata are especially
prone to undergo a change, whereby large tracts of
the tumor substance soften and become transformed
into mucin. When this takes place extensively
the tumor is converted into a spurious cyst. He
adds that the degeneration is preceded by edema
of the connective tissue and that the cells assume the
characteristic spider-like form to which the term
myxoma is applied. Virchow says this myxomatous
degeneration is characterized by an effusion of mucous
fluid among the muscular bands, and that it is distinguished from simple edema by the presence of
mucin and the multiplication of the nuclei and small
cells. When the bands between the small round cells
and the edema disappear, small fluid collections exist
which form the spurious fibrocysts.

The true fibrocysts, according to Pozzi, have a very

different origin. "These cysts are formed in pre-existing cavities, in dilated lymph spaces comparable to the similar dilatations which the blood vessels present. The fluid which they contain is limpid and coagulates on contact with the air. Leopold has termed these tumors 'lymph angiectatic myoma.' It must be noted that this lymphatic origin of certain cystic tumors of the uterus had already been clearly formulated by Koeberlé. This formation seems to be due to the development of part of the tumor along the path of the lymph vessels contained in the broad ligament. On the internal surface of such tumors we can demonstrate an epithelial investment which distinguishes them from simple cavities formed from softening of the neoplasm or apoplexy into its substance. There are also mixed forms in part vascular and in part lymphatic."

Induration.—Fibroids contract, reduce and harden as a rule following the menopause, while occasionally they disappear entirely. The same change takes place after confinement, in myomata of the fundus of the uterus which complicates pregnancy. The change resembles much that occurs in a fibroid after the removal of the appendages. The cause of the change, in each instance, is doubtless due, to a lessening of the blood supply.

Fatty Degeneration.—Literature does not record many well authenticated cases of fatty degeneration of fibroids. That it does occasionally occur, there is little doubt, although so rarely that from a clinical standpoint it is of little value.

Calcareous Degeneration.—Old uterine myomata, both large and small, are liable to become infiltrated with earthy matter. The change only occurs in slow growing tumors containing a large proportion of fibrous tissue. The calcareous material is not deposited in an irregular manner in the tissues of the tumor, but corresponds to the disposition of the fibers: on examining the sawn surface of a completely calcified uterine myoma we find the whorled disposition of the

fibers so completely reproduced as to leave no doubt
as to the nature of the mass. When these calcified
tumors are macerated and the decayed tissues washed
away, the earthy matter retains the shape of the tumor,
but its exterior presents an irregular, porous, almost
worm-eaten appearance. The calcification is confined
to the tumor itself, and though we may occasionally
find isolated nodules of earthy matter dotted about
the capsule, this part of the tumor is not converted
into a hard resisting shell.

Suppuration.—Suppurating fibroids may be fibroid
enlargements of any variety which by some accident
have become infected. Fortunately it is a condition
which occurs but rarely. Occasionally a subperitoneal
fibroid, which in process of pedunculation has been
deprived of a portion of its blood supply, by a gradual narrowing of the pedicle, may become infected by
migration of microbes through the walls of an adjacent
intestine upon which pressure has been long maintained. An interstitial fibroid may be converted, in
rare instances, into a suppurating mass through infection, by direct continuity, from a suppurating endometritis. Submucous polypi, of low vitality and small
resistance to pathogenic microbes, quite often are
infected from the endometrium and the vaginal secretions, and suppurate, resulting in an offensive discharge
of pus, and occasionally, hemorrhages—all of which
may give rise to a suspicion of malignant disease of
the uterus.

Carcinomatous Degeneration.—I have had an
opportunity of examining and having under observation, for long periods, an unusually large number of
fibroid tumors of the uterus, and I have never known
one to undergo carcinomatous degeneration. My
experience coincides with that of the best authorities
on this subject. Fibroid tumors do not predispose to
carcinomatous degeneration. Cancerous changes may
occur, however, in a fibroid uterus as a coincidental
disease, but in no way as a direct result of the fibroid.

Spontaneous Disappearance.—Almost every author-

ity who has watched the course of many fibroid
tumors of the uterus, has witnessed the spontaneous
disappearance of one or more of these growths, without any apparent cause. Gusserow at one time succeeded in gathering from literature thirty cases in
which this undoubtedly occurred. Of these thirty,
thirteen were associated with pregnancy, while the
majority of the remainder were connected with the
menopause. It is not difficult to explain the disappearance of a fibroid which has been coincidental with
pregnancy; the process of involution of the uterus
which occurs after confinement, is imparted to the
myoma, which so nearly resembles the uterine tissue,
and the tumor, consequently, is greatly diminished,
or disappears altogether. Then too, the decrease of
the blood supply to the uterus, as involution takes
place, deprives the tumor of its accustomed nourishment, and thus causes its diminution.

The reduction in size, or disappearance of a fibroid
at the menopause may be accounted for on the theory
of diminished blood supply, and consequent starvation. When senile atrophy begins in the organs of the
pelvis, and menstruation ceases, with consequent
decrease in the requirements of blood supply, the
uterus atrophies, and necessarily a tumor dependent
upon that organ for nourishment, must also suffer
anemia and reduction in growth.

But, occasionally, fibroids of the interstitial variety
will suddenly and mysteriously disappear, without
any apparent cause. I know of one which decreased
in size, fully two-thirds, as the result of a simple
exploratory operation.

These tumors may also be spontaneously expelled
by:

1. Pedunculation—a gradual narrowing and lengthening of the pedicle, until by violence or suppuration
the stem separates. This may occur with either a
subperitoneal or a submucous polypus.

2. Enucleation—This may occur with a submucous
or an intramural of the hard variety (fibromata)

which is usu illy surrounded with a loose capsule, and
which at its nearest approach to the mucous membrane becomes infected. The suppuration will gradually encircle the mass, and by means of uterine contraction, the tumor will slowly be shelled from its bed
and expelled from the uterus.

3. Suppuration.—Spontaneous disappearance of
fibroids occurs as the result of suppuration following
infection.

ETIOLOGY.

Pathologists have been unable to satisfactorily demonstrate the causes for the development of fibroids of
the uterus.

Pozzi says of Velpeau's theory, attributing the
development of fibroids to the presence of a blood
clot in the uterine tissue: "The spontaneous organization of coagula after ligation of the arteries suggested the idea that the same process might result in
the formation of these neoplasms. But experimental
study has demonstrated that this organization of coagula is nothing but an ingrowth of the elements of the
vessel's wall and thus this edifice of theory founded
on lack of observation, collapses altogether.

"Klebs asserts (Pozzi) that these tumors have their
origin in a proliferation of the connective tissue and
muscular layers of certain vessels; the different nodules thus formed become aggregated to make one
tumor. Kleinwachter describes the evolution of fibromata as due to a round cell which is found along the
capillaries and produces a partial obliteration of them:
these cells then become fusiform and produce nodules. In other words, our knowledge of the subject
is still very imperfect."

Winkle, without any clear demonstration or explanation, attributes the dependence of these growths
to the peculiarities of the vessels of the uterus, in
that the arteries are subjected to a very high pressure
before they reach the uterine wall, notwithstanding
their convoluted course

Senn believes in Cohnheim's theory of tumor development as modified by himself, viz.: That tumors never develop from mature tissue but from a matrix of embryonic tissue. According to Senn this matrix of embryonic tissue may be either of pre- or post-natal origin. "A fibroid," he says, "in the majority of cases springs, no doubt, from a matrix of mesoblast in the uterine tissue, while in exceptional cases the tumor may start from a similar matrix in the wall of blood vessels."

AGE AND DOMESTIC CONDITION AS PREDISPOSING
CAUSES.

Fibroids seldom occur before puberty. In 575 autopsies upon females by Winkle, 12 per cent. had these developments. Of 135 examined in the dead house by him, under 35 years of age, only 5 per cent. had fibroids.

Dr. Emmet, who has made a most careful study of this subject, basing his opinion on recorded cases, says: "It is impossible to determine with accuracy the age at which these growths are most likely to appear, since their development is, as a rule, slow at first, and they may exist for an indefinite period before their presence is recognized. The age can only be approximately inferred from the average one at which professional advice was first sought, and this would seldom be before the tumor had reached a sufficient size to cause hemorrhage or some other disturbance. We may also gain some information as to the rapidity of growth from the length of time elapsing after the birth of the last child, for a fibroid, it is well known is a cause of sterility. In the table is shown the age at which 225 women, who had fibroid growths, were first examined. The earliest age was 18, an unmarried woman; the next a sterile woman at the age of 22; 1 at 23; 10 between the ages of 24 and 25." His table then gives 25 cases between the ages of 25 and 30; 50 cases between the ages of 30 and 35; 48 cases between the ages of 35 and 40; 42 cases between

the ages of 40 and 45; 25 cases between the ages of
45 and 50; 8 cases between the ages of 50 and 55;
and 5 cases between the ages 55 and 60. Thus according to this table, which corresponds closely with the
experience of other writers, the age of greatest liability to fibroids is shown to be between 30 and 35
years.

According to Emmet, based upon statistical tables,
"between the ages of 30 and 40 years the unmarried
woman is fully twice as subject to fibrous tumors as
the sterile or as the fruitful;" he adds, "It seems as
if it were the purpose of nature that the uterus should
undergo the changes dependent upon pregnancy and
lactation about three years throughout the childbearing period, and that if the uterus is not physiolog-
ically occupied in childbearing, a fibroid will the
more rapidly develop. . . . This will also be the
case with the married woman who has taken means to
prevent conception, as well as with her who has been
sterile from some cause beyond her control, but to a
less degree in the latter case. . . . Finally, the
woman who may have been fruitful in early life, but
remained sterile long afterward from some accidental
cause, may have a tumor developed, but is less liable
thereto from having once borne a child."

Brooks Wells says: "Myomata of the uterus are
more common in old maids than in married women."
This statement is often disputed by gynecologists who
do not frequent the dead house. Very many examples of myomata appear postmortem whose presence
was not even suspected during life.

Race.—It is an undeniable fact that the negro
women, with the environments of this country, are
more liable to have fibroids of the uterus than
white women. This statement is doubted by Dr.
Middleton Mitchel, of Charleston, S. C. I can not
but believe, however, that Mitchel is wrong, especially as far as the negro women living in the
northern and colder latitudes of the country are
concerned.

16

SUMMARY.

The cause of fibroid tumors of the uterus has never been satisfactorily demonstrated.

Fibroids of the uterus rarely occur before puberty, and seldom before the age of 25, while the greatest number develop between the ages of 30 and 40 years.

The unmarried state predisposes women to the development of uterine fibroids. Married women who prevent conception, while less liable to develop fibroids than unmarried women, are still much more prone to them than childbearing women. Childbearing women are the least predisposed to fibroid of the uterus.

Negro women are predisposed to uterine fibroids.

LECTURE II.

A knowledge of the existence of a fibroid tumor of the uterus may be gained by the diagnostician by first obtaining the symptoms as appreciated by the patient—the subjective symptoms; and further by acquainting himself with the actual physical changes by direct personal examination of the patient—the objective symptoms.

SUBJECTIVE SYMPTOMS.

Pelvic Symptoms.—Among the early local symptoms of fibroid tumors of the uterus may be enumerated an irritable bladder amounting frequently to positive dysurea; rectal pressure; sensation of pelvic fullness; low backache or sacralgia, and frequently pain on cohabitation. These symptoms are all produced by a gradually enlarging uterus, and resemble many of the pelvic disturbances of early pregnancy, from which they must be differentiated. As the tumor enlarges the sensation of fullness extends to the lower abdomen, the pressure on the nerves to the lower extremities causes pain in the line of the nerves on the anterior or posterior aspect of the thigh, or on both. Even edema of the extremities may occur from pressure on the veins extending to them, and the appendages are frequently pressed upon, resulting in severe pain on one or both sides, while as the tumor begins to fill the abdomen symptoms of painful pressure on many or all the important organs of the pelvis will be experienced.

Symptoms due to Functional Disturbances.—The
most important symptom under this head is that due
to the disturbance of the function of menstruation.
In 75 per cent. of all fibroid tumors of the uterus the
menstrual flow is increased, on account, 1, of increased
area of the endometrium due to interstitial enlargement of the uterus, 2, of increased vascularity of the
uterus due to the demands of the hypertrophied tissues, and 3, to the venous blood congestion due to
pelvic pressure. In a large majority of cases pain is
an accompanying symptom. This dysmenorrhea is
caused, either by the abnormal contractile power of
the changed uterus, by submucous projections into
the uterine cavity, exciting painful contractions of the
organ, or by a frequently accompanying endometritis.
The development of the tumor influences decidedly
the change of the menstrual function. In the early
stages of the fibroid the patient will notice but a slight
lengthening of the menstruation, but as further development is made the quantity of the flow will be increased. This changed condition, while at first it may
attract the attention very slightly, as it gradually increases, will at last convince the patient that something serious is afflicting her. The flow will increase
rapidly, not only in length of period, and in quantity
at a given time, but finally it will frequently become
irregular and occasionally almost continuous. At the
same time pain will often gradually develop, so that
with the exhaustion of depletion will come the agony
of physical suffering. These pains if caused by endometritis, will be of a dull aching character, accompanied occasionally with slight uterine contractions;
if caused by the effort of the uterus to expel submucous masses or polypi, it will be like those accompanying the uterine expulsive pains of a miscarriage or
confinement; if caused by a pressure of the inordin-
ately congested hypertrophied uterus upon the tubes
and ovaries it will be severe and of an almost continuous character in the ovarian regions. ·

The function of the bladder suffers from direct

pressure of the enlarged uterus, or from a subperito-neal enlargement. Frequent urination will first be noticeable, while the tumor is yet small, and later pain-ful micturition, with severe lasting pain in the bladder, as a direct result of traumatism produced by the en-croachments of the uterus. Complete stoppage of the urine and painful distention of the bladder may finally occur from impaction of the increasing tumor.

The function of the rectum is frequently impaired by direct pressure of the tumor upon that organ. Obstinate constipation will be complained of, while the interference of the tumor with the circulation will favor the development of hemorrhoids and their pain-ful symptoms. Temporary impaction of feces in the large and small intestines occur as a direct result of the pressure of a large tumor.

Deformity Produced by Fibroids. One of the most embarrassing symptoms to many patients, who are afflicted with fibroid tumors, and frequently the first to attract their attention, is the change in the contour of the abdomen, which is enlarging. Upon closer ob-servation and examination of the lower abdomen, they discover the unwieldly mass, the tumor, which, as it gradually increases in size, produces a deformity that no device can conceal, while the patient, in order to maintain an upright position, must throw her shoul-ders back in a manner to make the tumor appear most embarrassingly conspicuous.

General Constitutional Disturbances.—Reflex ner-vous disturbances are early symptoms in many of these cases. Nausea, palpitation of the heart, indi-gestion, gaseous distensions of the bowels, flashes of heat due to vasomotor disturbances, headache, dizzi-ness, occasionally spasmodic cough and all the symp-toms accompanying nervous storms, irritable temper, despondency, and lack of control, finally developing into typical hysteria. These symptoms are precipi-tated ordinarily by irritation caused by the long-con-tinued local disturbances already described, and by loss of blood due to excessive menstruation.

Anemia naturally, in a large number of cases, becomes a conspicuous subjective symptom. The patient will complain of loss of flesh, her skin has become pale, lips pale and blue, muscles loose and flabby. She tires easily, and all exercise induces shortness of breath and heart palpitations. Frequently it is necessary to remain in bed a large part of the time from weakness, which is much more pronounced immediately following the great waste of the menstrual period.

OBJECTIVE SYMPTOMS.

Diagnosis.— Under objective symptoms methods of diagnosis will be considered. Objective symptoms are determined, and their diagnostic value recognized, by pelvic examination, abdominal palpation and auscultation, and general examination.

Pelvic Examination.— The patient should be placed on her back on an operating chair or table, in the recumbent position, with limbs well flexed upon the abdomen, and feet supported by short stirrups. The clothing should be loose. After the vagina has been well douched with an antiseptic fluid, the examiner should proceed to make a bimanual pelvic examination. The index finger of the left hand is employed to make the preliminary vaginal examination, the right hand being left free for external manipulation. The cervix is first sought and its location often gives one important information. If it is in normal position, well up in the vagina, within an inch and a half of the sacrum, any enlargement of the uterus is liable to be uniform, or any growth is located anteriorly in the body of the organ. If it is low in the pelvis and anterior to its natural location, any enlargement of the uterus is quite likely to exist in the fundus, or the uterus is retroverted. If it is anterior to its normal position and is crowded well up behind the symphysis, the tumor is liable to be located in the posterior wall of the uterus, very low. If it is drawn well up posteriorly, almost if not quite beyond reach, the

tumor will usually be found in the anterior uterine
wall. The location of the cervix, however, is of small
importance when considered alone.

Bimanual Examination.—By combining with the
digito-vaginal examination, external palpation with
the hand, a knowledge of the general contour of the
uterus can be at once definitely obtained. If it is enlarged, that fact is apparent; if there are any irregular
projections or developments from any portion of its
walls into the abdominal cavity, or into the broad ligaments, they can be recognized; and if it is enlarged
so as to produce a prominent tumor a knowledge of its
source of development can usually be determined by
taking into consideration its relation to the cervix. If
the tumor is of a subperitoneal variety, its relation
with the cervix and uterine body will indicate its
source of development. If the cervix is thin and
stretched over a projecting mass, protruding through
it from the uterine cavity, the knowledge of a uterine
polypus or a submucous tumor is imparted. The sensation of solidity or fluctuation of the growth is usually
satisfactorily obtained by this examination. Thus,
the size, contour, consistence, direction of development and variety of tumor may all be determined by a
simple bimanual pelvic examination.

A bimanual examination, however, is frequently
unsatisfactory, until one has resorted to a rectal examination, combined with the abdominal palpation, and
made either with or without an anesthetic. This latter enables one to examine the posterior aspect of the
uterus much higher than is possible per vaginam, and
with less intervening tissue. An anesthetic, when
given, permits of a much more prolonged examination,
renders the procedure painless, and eliminates the
muscular fixation and strain which is inseparable from
an examination without it. The abdominal muscles,
the sphincters of the rectum and vagina are relaxed,
allowing much freer and more complete exploration,
when an anesthetic is employed.

Instrumental Examination.—Pelvic examination

for fibroids is often not complete without instrumental examination. While a speculum is not often needed as a direct means of diagnosis, in these cases, it is frequently required to aid in determining secondary changes which may have occurred in the vagina or cervix uteri. It may also be employed to explore the cervix where the vaginal vault has been distorted by a complicated tumor, before using the uterine sound. The uterine sound is occasionally used to measure the uterine canal. It may be employed to locate the canal when palpating the uterus for the purpose of ascertaining in which portion of its mass a tumor is situated. And it will also give valuable information in regard to the location and size of submucous fibroids. When a uterine canal is so distorted that a metal sound does not possess the necessary flexibility to traverse its course, an ordinary male flexible bougie, with a bulbous tip, can frequently be successfully insinuated to its entire depth, and give very valuable aid in diagnosis. The uterine dilator is an important instrument in diagnosing the condition of the interior of the uterus. It enables one to explore the cavity of the uterus with the finger and thereby materially assists in revealing the position and character of submucous fibroids.

Abdominal Palpation.—Abdominal palpation and auscultation can best be accomplished with the patient under the influence of an anesthetic. By palpation the tumor can be outlined; any subperitoneal projections noted; its consistency ascertained; abscess or presence of fluctuation considered; its relations to the surrounding organs recognized and the absence or presence of adhesions learned. In auscultation we have an important means of distinguishing fetal heart sounds in the differential diagnosis between a fibroid tumor and a natural or tubal pregnancy.

General Examination.—In the general examination of a patient who has a fibroid tumor, points of change in outlying organs, which may have been the remote result of the tumor and which may have an important bearing on the form of treatment to be adopted, are to be noted. These changes may be found in the heart, the kidneys, the lungs, the digestive tract, the nervous system and the vascular system; but as similar changes may also occur from any form of wasting disease, as diagnostic signs they are not of great importance, and therefore, their consideration here is out of place.

DIFFERENTIAL DIAGNOSIS.

We are often called upon to differentiate between fibroids and the following conditions, which simulate them to a puzzling degree: Ante- and retro-flexions, subinvolution (metritis), cancer, floating kidney, pregnancy, tubal pregnancy, tubal cysts and ovarian cysts.

Ante- and retro-flexions are differentiated from fibroids by the comparative smallness of the uterus, and the lack of characteristic hardness of the fibroid. The sound clears up the condition. If it is that of a normal uterus with projecting fibroid, the mass will be easily palpated in front or posterior to the sound, while if it is ante- or retro-version the sound will follow the canal to the center of the flexed fundus.

Subinvolution (metritis) is distinguished from a fibroid by its history, by its symptoms and by physical signs. The history of subinvolution is that of recent child bearing or miscarriages with subsequent endometritis, while a fibroid seldom immediately follows pregnancy, and when small is rarely accompanied with endometritis. The symptoms of the two conditions differ in the amount and character of the leucorrhea, being profuse in chronic metritis; in the amount of hemorrhage at menstruation, being profuse in fibroids, while in subinvolution it may be absent immediately following child birth for several months, and of a more normal character when it does exist. Subinvolution in physical signs differs from a fibroid in being of a softer character, resembling the normal uterus in being of uniform contour and in containing no distinct centers of development. With a condition of subinvolution will almost invariably be found an accompanying lacerated cervix and frequently a lacerated perineum. Finally, uterine curettement, antiseptics and a restoration of a torn cervix, will cause a subinvoluted uterus to involute to its natural size, while such will not be the case with a similar treatment for a fibroid.

A cancer (uncomplicated) is usually easy to differentiate from a fibroid. Microscopic examination should be made, if practicable, at the earliest possible date. An uncomplicated carcinoma seldom presents a large tumor. If it is of the cervix, a well-defined ring shows the line of demarkation between it and the healthy tissue. An offensive, watery discharge, characteristic of carcinomatous degeneration, is of diagnostic value. Carcinoma of the body of the uterus gives rise to more pain than a fibroma. It, too, ulcerates early, giving rise to the afore-mentioned offensive discharge. The cancer cachexia will also aid in arriving at a differentiation. A carcinomatous tumor is much more liable to give rise to an ascites than a fibroma.

A floating kidney might, under exceptional circumstances, be mistaken for a subperitoneal fibroid. But its position, and especially the discovery of its place of attachment, would dispel all doubt of its being a fibroid. To obtain this knowledge, however, might require a most careful abdominal and bi-manual manipulation.

Normal pregnancy should be differentiated from fibroid tumors first by bearing in mind the subjective symptoms of the two conditions, and second by physical examination. In pregnancy menstruation almost invariably ceases; in fibroids it is almost invariably increased. The history of pregnancy is definite and uniform, with a given size of tumor, while with fibroids the time of growth varies greatly with tumors of the same size. Morning sickness and breast changes are classic symptoms of pregnancy which rarely occur with fibroids. In the physical examination the pregnant uterus is uniform in its development, giving a soft, semi-fluctuation feel on palpation. Fibroids are frequently irregular in outline, and on palpation appear solid and often present distinct centers of hardness. The cervix is soft and patulous in pregnancy, and the neck of the uterus sometimes thin, forming a decided constriction, while in fibroids the cervix is unyielding and the constriction of the neck is absent. The vagina in pregnancy is blue, a rare occurrence in fibroids. Finally, as the pregnancy progresses ballottement, fetal movements and fetal heart sounds remove all doubt of the condition.

Tubal pregnancy would rarely be mistaken for a fibroid of the uterus. It presents many of the signs of normal pregnancy. There is frequently a slight show at menstruation, the tumor is of a semi-fluctuating character, and lies, as a rule, to the right or left of the uterus; while a subperitoneal fibroid, located in the position usually occupied by a tubal pregnancy, would be hard and unyielding. In tubal pregnancy the normal uterus may be distinguished by a line of demarkation between it and the tumor.

Ovarian cysts are usually easy to distinguish from fibroids by their fluctuation; by the lateral delopment when small; by the slight menstruation accompanying them; by the normal size and depth of the uterus and by the absence of the uterine souffle distinguishing a fibroid. As a rule, ovarian cysts are of a rapid growth and give a short history, while fibroids are years in maturing.

LECTURE III.

TREATMENT—GENERAL CONSIDERATION—MEDICAL TREATMENT.

In the consideration of the histology, growth, anatomy, varieties and symptomatology of fibroids of the uterus one is irresistibly drawn to the conclusion that each one of these growths is a law unto itself. Such is certainly the case when we come to the consideration of treatment.

A fibroid tumor is a benign growth. The principal symptoms arising from it are due to the effect of its presence upon organs associated with it or in close proximity to it rather than from any inherent qualities. From its slowness, in the majority of instances, to produce serious symptoms, and the rarity of a fatal termination as a direct result of it, patients are with difficulty aroused, early in the history of the disease, to adopt radical measures for relief. Hence innumerable symptomatic remedies have arisen, each for a brief time having enjoyed the reputation of specifics, only to sink into their true positions as soon as the light of experience had shone upon them. The simple enumeration of a list of these remedies would fill a long chapter.

The remedies for fibroid of the uterus which merit attention must be considered under three principal heads:

1. Medical treatment.
2. Electrical treatment.
3. Surgical treatment.

To indicate when each of the above methods of treatment could be employed with the greatest advant-

age to the exclusion of the other, how or when two or
more could be combined to greatest advantage would
be a task which would necessitate the careful analysis
of a large number of fibroid cases, together with all
the peculiarities of the growths with their surround-
ings, the temperament and the physical condition of
each particular patient possessing the tumor. The
most that can be attempted is to describe the various
remedies; the exact physiologic or mechanical effects
of each; the manner of administering drugs; the
chemic, mechanical and physical effects of electricity,
with a comprehensive description of the details of application; the many operative procedures including
the rationale of each with a minute technical description of those which merit attention; and finally to
show wherever each remedy is especially adapted to
counteract some specific condition or symptom of a
fibroid uterus.

I will therefore consider each grand division of
treatment in detail trusting to the diagnostician to apply the remedies as the indications in each particular
case may dictate.

MEDICAL TREATMENT.

The medical treatment of fibroids of the uterus has
played an important part in the symptomatic relief
and the prolongation of life of those afflicted with this
dread condition. For convenience its therapeutics
may be divided into general tonics, alteratives, sedatives, astringents and special uterine tonics.

General Tonics.—Rapid growth of a fibroid tumor
and a condition of general good health in a patient seldom exist together. So true is this, that I
have learned to assure patients laboring under this
disease that their tumors will not increase in size rapidly if they can by some means keep themselves in a
condition of natural tonicity. While undoubtedly the
cause, frequently, of the general lack of tone is the
rapidly growing tumor, I believe its growth may be
retarded by adopting means to counteract its effect

upon the patient. While the tumor with its accompanying pain and loss of blood, and rapid growth must impair general health, that impaired tone of necessity leaves the patient with still less resistance to withstand the onward march of the disease. In other words the results of the rapid growth of the tumor does in turn become one of its aids to still more rapid destruction. Therefore fibroid-tumor patients should be abundantly nourished. The loss of blood should be met by an abundant supply of wholesome substantial food. Bitter tonics such as quassia, calumbo, nux vomica and cinchona should be administered to increase the appetite. Iron should be administered in a form to be readily assimilated in order to improve the blood. The patient should be urged to drink milk and cream between the regular meals. Prepared foods both of the nitrogenous and farinaceous order should be employed to reinforce the ordinary diet. Predigested foods may be found of service if the digestive organs fail to stand the strain of over feeding.

As nerve tonics the compound syrup of the hypophosites with small doses of quinin, strichnia and iron are often of great service. Strychnia in gradually increased doses until the patient takes 1-10 of a grain four times a day, I consider of great value as a tonic in these cases. The dose should begin at 1-40 or 1-60 and be increased gradually, being careful to avoid unpleasant constitutional symptoms. This tones the bowels, acts as a vaso-constrictor tonic everywhere, and I believe acts as a direct tonic to the muscular structure of the uterus itself, while it undoubtedly acts as a vaso-constrictor to the uterus and tumor. As a direct bowel stimulant and laxative aloin with strychnia and hyoscyamus has few equals. Cascara sagrada may take the place of the above as an alternative laxative. General stimulating baths often prove beneficial tonics to fibroid patients. Sea salt baths, warm or cold with judicious rubbing to establish reaction, are often recommended. Chlorid of sodium baths may be employed in lieu of the sea salt baths.

Hot water as a local tonic, employed as a vaginal douche, frequently relieves excessive congestion of the uterus. To accomplish this object the water must exceed a temperature of 115 F., the patient should use at least two gallons at a time and should employ it while in the recumbent position on the back. Chlorid of sodium as an antiseptic and tonic may be dissolved in the water—3 to the pint. Or to get the advantage of an astringent the same amount of powdered alum will prove of advantage. The water must be hot. Tepid water will favor hemorrhage.

Change of air from the sea shore to the mountains (not too high an altitude) or from the mountains to the sea shore often improves health and retards the growth of these mysterious tumors. An ocean or a European trip will often bring a remarkable change for the better in these unfortunate cases. Even a slight move which only involves a change of routine, or surrounding the patient with a pleasant change of companions will draw them away from themselves and occasionally give them just the impulse which they require to regain the lost tone.

Alteratives.—Arsenic, muriate of ammonia, mercury and iodin have an important role in the therapeutics of uterine fibroids.

Arsenic in the form of the acid or in Fowler's solution given in 1-20 dose of the former and in 5 drop doses of the latter immediately following each meal will often act as a powerful alterative tonic in these cases. While no specific action on the growth of the tumor has been demonstrated, its general effects are frequently so satisfactory in maintaining a general condition of tone that it is strange it has not run the gauntlet of a specific.

Muriate of ammonia, which has none of the tonic properties of arsenic and while inferior in alterative properties, has enjoyed a brief trial as a specific for fibroids. Its principle action, it seems to me, lies in the direction of a hepatic stimulant. By favoring a free portal circulation its action in relieving uterine

and pelvic congestion must be obvious. I frequently
employ it in 5 to 10 grain doses to be taken in one-
half glass of water on rising in the morning. It acts
in this way as a very gentle stimulant to the bowels
and may be used as a substitute for saline mineral
water. As a specific for fibroids, however, one must
not place upon it much reliance.

Mercury in the form of blue mass or calomel, is the
most valuable alterative, laxative and tonic we have in
all pelvic difficulties. Its efficiency and superiority
as a laxative lies in the fact that it is our best hepatic
stimulant. It unloads the portal circulation, relieves
pelvic congestions and favors the rapid removal of
waste products from the pelvic organs. It too, possibly from the same reason, increases the patient's
desire for food, aids healthy digestion, favors normal
assimilation and gives an impetus to the general circulation. My esteemed friend and teacher, the late
Prof. W. H. Byford, taught me the value of this
remedy as a tonic, when properly administered, in
gynecologic practice. As a laxative the blue mass
should be given in 1 to 2 grain doses every evening
for several days, then to be omitted for a like number
of days. Following the night dose the next morning
should be administered a mild saline laxative, either
a seidlitz powder, or from a drachm to two drachms of
granulated citrate of magnesia in one-half glass of
water. As an alterative tonic it may be given in half
the above doses extending over a period of twelve days.
Calomel is a convenient form for administering mercury as a tonic and alterative. Triturated with sugar
of milk, doses of 1-30 of a grain twice or three times a
day may be given for a period of twelve days. An
interval of twelve days should then elapse before re-
continuing the drug. The bowels should be kept free
with saline laxative while the mercurials are in use,
and care should be exercised to avoid the unpleasant
consequences of administering hydrochloric acid co-
incident with the drug.

Iodin in the form of tincture applied to the cervix

of a fibroid uterus or employed as an application to the mucous membrane has been much extolled as a routine treatment for fibroids. As an alterative its effects on a fibroid tumor must be slight, but as a stimulant and antiseptic application to the endometrium of an endometritic uterus it certainly will do good. The injection of iodin into the uterine tumors will scarcely be tolereted as a rational or at least as a desirable treatment at the present day.

Sedatives. The bromids, the coal tar antipyretics, chloral, cannabis indica, hyoscyamus and opium are sedatives which succeed in relieving the suffering of many fibroid patients and which, when judiciously employed succeed occasionally in tiding them over desperate attacks which would prove fatal without their assistance. In chronic conditions like this one we are considering, no physician should mistake the temporary effects of these sedatives for effects of a curative agent. Symptomatic relief only can we hope from them while we are selecting and marshaling our final conquering remedies.

The bromids are probably as safe and as enduring as any mild sedative, where irritability and general nervousness characterize the patient's suffering instead of actual localized pain. Bromid of potassium, ammonia and sodium mixed in solution given in doses containing 10 grains of each, the doses given at intervals of four to six hours, until nervousness is allayed and sleep obtained, is a favorite method of mine of administering this drug. Frequently the above combination given as a dose at bedtime will accomplish good.

Antifebrin, phenacetin and antipyrin I have found of value in relieving severe menstrual pain accompanying fibroids. By giving them in their appropriate doses at intervals of two to four hours during the painful period of the menstruation frequently complete relief is obtained.

Chloral is a favorite hospital remedy of mine. I employ it for the uterine pain which accompanies

menstruation exclusively—not as a steady remedy
The favorite method of administering it is to instruct
the nurse to give 20 to 30 grains in three ounces of
water as a small rectal enema not oftener than once in
four hours for severe pain or sleeplessness. It does
not affect the stomach when given in this manner,
and the method of administering the remedy serves to
make it unpopular to the patient. In fact, an effort
is made to keep the patient in ignorance of the contents of the enema.

Cannabis indica as a remedy for pain in menstrual
difficulties is one which I have never learned to
appreciate. The remedy is so uncertain and variable
in its actions that I usually resort to something upon
which I have greater reliance. If a pure article can
be obtained a pleasant combination of cannabis indica
in extract combined with ergot and valerinate of quinin for the pain of fibroid menorrhagia may be
employed with advantage. The extract of cannabis
indica in from $\frac{1}{4}$ to $\frac{3}{4}$ of a grain, the ergotin in from $\frac{1}{2}$
to 1 grain, and the valerinate of quinin in 2 grain
doses, makes a powerful uterine tonic, the cannabis
indica and the valerinate modifying the pain which
otherwise would be aggravated by the necessary
effects of ergot and quinin.

Hyoscyamus, belladonna and stramonium are valuable sedative remedies to employ as alternatives in the
long medical treatment which one is often obliged to
conduct before these patients decide to have radical
means adopted. While one sedative, hypnotic or anti-
irritant remedy after another is being employed these
drugs will often give surprising aid as a temporary
substitute for remedies which for prudential reasons
we feel should be discontinued. The active principle
hyoscyamin or atropin may be given in solution in
appropriate doses; or may be given in the small gran-
ules which are so accurately compounded by the large
manufacturers of pills. I frequently combine these
drugs with the valerinate, the two making very effec-
tual sedatives or anti-nervous remedies.

Gelsemium, valerian, and asafetida are very valuable drugs, which can be employed with success in the
treatment of fibroids. Extract of gelsemium, combined with valerinate of quinin, zinc and iron, acts
as a very reliable aid as a mild anti-neuralgic, while
in asafetida we have a remedy which the physician
could ill afford to lose. Good full doses of asafetida,
4 to 8 grains, four times a day has carried many a
nervous, sleepless, hysterical woman successfully
through what otherwise would have been a painful
menstrual period.

Opium is a dangerous remedy to employ for the
pain of fibroids. The remedy is so prompt and efficient with the majority of patients that it is an easy
matter for a dangerous habit to be contracted, inas-
much as the difficulty is a chronic and constantly
reverting one. If in desperate cases it is considered
wise to employ the drug temporarily, it should be
presented by the physician and given, if possible,
without the patient becoming aware of the contents
of the dose. A condition serious enough, however, to
demand the use of opium should also demand some
permanent radical means of cure.

Astringents.—Local astringents include such drugs
as may be employed with curative benefit by application to the mucous membrane of the uterus or by application to the neck of the womb through the vagina.
Local astringents may be employed, too, for the temporary effect they may have on the secretions of the
uterus or for the direct curative effect they may exert
when applied to the mucous membrane of the fibroid
womb. When the astringents are applied to modify
temporarily the secretions of the uterus, for instance
the excessive blood flow, it is best applied on tampons.
Alum is cheap and one of the safest and most efficient
astringents to use in this way. Wicking, or strips of
gauze, or cotton tampons may be prepared and kept
for emergencies, by dipping them into a 20 per cent.
solution of alum in hot water and then evaporating
the water of solution by exposure to the air. This

leaves the alum thickly deposited in the interstices of the material. It is only necessary then to apply the tampon directly to the bleeding cavity, allowing the fluids of the cavity to dissolve the alum, or it can be placed in warm water immediately before applying. Excessive bleeding from the uterus which fails to yield to other remedies will seldom fail to be controlled by well placed uterine tampons well saturated with alum. Wicking or strips of gauze without knots is the best material to employ. The patient should be placed in Sims' position the neck of the uterus grasped with a strong tenaculum, so as to expose the canal, then by means of a bulbous-pointed uterine sound, the material should be systematically packed to the bottom of the uterus. Only where the hemorrhage is alarming is this procedure necessary. In ordinary bleeding a well-placed vaginal tampon alone is necessary. For vaginal tampons, Byford recommends compressed sponges saturated with alum. They are made by taking a fine sponge, large enough to fill the vagina, passing a strong string through the center to aid in its removal, and then after dipping it in the solution of alum, well winding it with twine from one end to the other, compressing it into as small a space as possible. The twine should so compress the sponge as to make it assume an elongated form. It should then be laid aside and permitted to dry. Several of these should be kept on hand by the patient, she having been previously instructed how to employ them. After compression and drying it will be of small size, and such shape that it can be easily introduced into the vagina, while the moisture of the parts will soon cause it to expand into an efficient tampon.

Astringents when applied to the mucous membrane of the uterus, are applied ordinarily for the curative effect they may exert on the inflamed endometrium. For this purpose the tincture of the chlorid of iron is a popular astringent. I have employed with benefit a 10 per cent. solution of chlorid of zinc in glycerin and water. Varying strength solutions of nitrate of silver have been employed. This is objectionable, however, because of the indelible stain produced by it. There is such a large variety of well-known astringents which may be employed for this purpose that it seems like needless waste of time to enumerate them.

The application to the mucous membrane should be made, after the cervix has been exposed, by means of a flexible applicator wrapped with cotton. The canal should first be wiped free as possible of all secretion by means of the applicator and cotton and when that is accomplished the mucous membrane should be thoroughly painted with the astringent by passing the applicator, well wound with absorbent cotton, and saturated with the drug, to the bottom of the uterus.

General Astringents.—Of the astringents which reach the uterus and modify uterine hemorrhage when given internally I know of none which compares with hydrastis canadensis. It contracts the blood vessels and modifies the hemorrhage. It is so efficient that one is led to believe that it has a direct contractile effect upon the muscular walls of the uterus and tumor. While it acts as an astringent with special predilection for the uterus it also is efficient as a bitter tonic. The fluid extract may be given in doses of a third to one-half a drahm three or four times a day. I frequently combine the solid extract with ergot in grain doses of each in form of capsules, one three or four times a day. This drug actually seems to possess a curative effect not only in diminishing the amount of the blood discharge but also in decreasing the size of the tumor. When employed to modify hemorrhage it should be given in good maximum doses of one-half to three-fourths drachm of the fluid extract three or four times a day during menstruation. If it is to be employed as a direct curative agent it can be more conveniently and elegantly administered in capsule form as already indicated either alone or combined with other uterine tonics.

Tincture of cannabis indica in from four to ten drop doses given during the approach and continuance of menstruation not only modifies irritation but also acts as an astringent and hemastatic.

Special uterine tonics.— Ergot has occupied such a conspicuous and important role in the treatment of fibroids of the uterus that I deem it advisable to devote a special article to its consideration.

The foregoing article can be briefly summarized as follows:

1. The medical therapeutics of fibroids of the uterus are divided into general tonics, alteratives, sedatives, astringents and special uterine tonics.

2. General tonics: Judicious and systematic feeding: bitter tonics including quassia, calumba, nux vomica and cinchona; iron; prepared foods of all kinds. Nerve tonics: The hypophosphites: strichnia and manganese. Bowel tonics: Aloin and cascara sagrada. General stimulants: Hot salt baths: sea salt baths. warm or cool: massage. Local tonics: Hot vaginal douches. Climate: Change of residence, travel and ocean voyage.

3. Alteratives: Arsenic; muriate of ammonia; mercury: iodin.

4. Sedatives: The bromids: the coal tar antipyretics; chloral: asafetida: valerian: gelsemium; hyoscyamus; opium.

5. Astringents Local. alum: chlorid of zinc: preparation of iron. General astringents: Hydrastis: cannabis indica.

6 Special uterine tonic: Ergot.

LECTURE IV.

ERGOT.

The physiologic actions of ergot are accounted for by its effects upon unstriped muscular fiber. It contracts blood vessels everywhere, thereby increasing the blood tension. The uterus, formed as it is by an enormous proportion of smooth muscular structure, is particularly susceptible to the peculiar action of ergot. It acts upon the uterus in four ways:

1. It decreases the whole bulk of the organ by producing a steady tonic contraction of all its muscular fibers; 2, it decreases the whole bulk of the organ by decreasing the amount of blood in its walls, as a result of contracting its blood vessels; 3, by decreasing the amount of blood in the uterus (by contracting its muscular bulk and by contracting its blood vessels) it modifies materially its nutrition and decreases the amount of the menstrual flow of blood; 4, when given in large doses it not only produces tonic contractions of the muscular fibers of the uterus, but, by instituting clonic contraction of its fibers, expulsion of natural and foreign bodies from its walls and cavity is effected.

Upon these thoroughly demonstrated effects of ergot has grown up a rational and very successful medical treatment of fibroids of the uterus. Before surgery and electricity found favor in the experience of gynecologists, as remedies which did their work with greater expedition, if not with more certain results, ergot for many years was depended upon, not only as a symptomatic remedy for fibroids but as a valuable curative agent.

The mode of action of ergot on fibroids of the uterus when considered in the light of its effect upon the normal uterus, can be easily understood.

An interstitial fibroid is affected by ergot in three ways: First, by the pressure exerted on the uniformly distributed fibro-myomatous fibers by the normal uterine tissue which everywhere surrounds them which has been stimulated to firm contraction by the drug; second, by a tendency of the hypertrophied fibro-myomatous fibers themselves to firmly contract under the abnormal stimulant. The result of this pressure upon the abnormal tissue is to produce atrophy. A third effect of ergot upon an interstitial fibroid of the uterus which favors its atrophy and decreases uterine hemorrhage, is the diminution of its blood supply from the contraction of its arterioles.

Intramural fibroids are also affected in three ways: 1, pressure atrophy from the firm contractions of the normal uterine tissue surrounding the masses; 2, atrophy and decrease of uterine hemorrhage, diminution of their blood supply as a direct result of contraction of the arterioles; 3, a tendency for them to be expelled from the walls or the uterus in the direction of least resistance, as a result of the persistent contraction of the normal uterine tissue surrounding the masses.

Submucous fibroids, which are but a form of the intramural variety, are influenced in the same manner by ergot as is the simple intramural, while its cure is also sometimes effected by the expulsive contractions of the uterus, which force it first beneath the mucous membrane as a pedunculated mass, and finally expel it bodily from its cavity.

Subperitoneal fibroids, which also are but a form of the intramural variety, are influenced practically the same as that class, except that the influence of muscular contraction diminishes as the tumors become more subperitoneal and more pedunculated. The

39

expulsion of the subperitoneal variety is not aided by
the expulsive power of the uterus.

INDICATIONS FOR ERGOTIN TREATMENT.

The most favorable case for ergotin treatment in
which a complete cure may be looked for is the intramural variety which consists of one or few centers of
development, with these centers situated in close
proximity to the mucous membrane. The uterus in
such a case is considerably enlarged, the canal is long
and tortuous, the symptom of uterine hemorrhage is
conspicuous and the normal muscular tissue of the
uterus is hypertrophied. Ergot given in such a case
in large doses, continued for some weeks, will cause
the hypertrophied muscles of the uterus to contract
and to gradually squeeze the fibromatous center first
toward the mucous membrane, then beneath the
mucous membrane and finally into the uterine cavity,
where the uterine expulsive pains will effect its deliv-
erance. The enucleation of such a tumor can be
materially assisted, as it should be, by incising its
capsule as it begins to protrude beneath the mucous
membrane.

When such a center or centers are expelled effectu-
ally the uterine pains will cease, the hemorrhage will
lessen and the uterus will soon be found of normal
size.

Pediculated fibroids are somewhat differently affected by ergot when they are submucous than when
subperitoneal. If the uterus is considerably enlarged
and the normal muscular tissue of the organ is hypertrophied, there is some hope that the pedicle of a subperitoneal pediculated fibroid will gradually become
restricted by the uterine contractions, the nourishment of the tumor supplied by the pedicle be diminished thereby, and as a consequence a gradual atrophy
of the stranded mass. While this *may* occur with a
subperitoneal tumor, it may be counted on much more
surely if the tumor is submucous.

In the *submucous pediculated* fibroid the nourish-

sive effort of the uterus itself will, in its effort to rid
itself of the foreign body, stretch and thin its pedicle
until its blood vessels are eradicated and its tumor
decreases from starvation. Sometimes the decrease is
only partial from some small amount of blood still
gaining access to the mass. Under either circumstances it remains but to catch the growth in a forceps
and to twist it from its weak stem. The uterus then
contracts to its normal state if there are no other centers of development.

A true interstitial fibroid is seldom, if ever, cured
by ergot. While the muscular fibers of the uterus
are often enormously developed, there are no distinct
masses for them to work on when stimulated to contraction; nothing for them to get rid of but their
fellow fibers. In this they may partially succeed by
producing atrophy of one another by pressure, and
by lessening the blood supply by contracting their
blood vessels. This, however, experience has taught
us is a slow method of diminishing a fibroid tumor.
If large doses are systematically administered for a
long time menorrhagia can often be materially benefited, but seldom permanently relieved.

Prof. W. H. Byford, who was the first in this country to publish results from the treatment of fibroid
tumors by ergot, after Hilderbrandt had published his
work in this line in Europe, begins his chapter on the
subject as follows: "1. When properly administered,
ergot frequently greatly ameliorates some of the
troublesome and even dangerous symptoms of fibrous
tumors of the uterus, e. g., hemorrhage and copious
leucorrhea. 2. It often arrests their growth and checks
hemorrhage. 3. In many instances it causes the absorption of the tumor, occasionally without giving the
patient any inconvenience; at other times removal of
the tumor by absorption is attended by painful contractions and tenderness of the uterus. 4. By induc-
ing uterine contraction it causes the expulsion of the

polypoid variety. 5. In the same way it causes the disruption and discharge of the submucous tumor."

METHOD OF ADMINISTRATION.

According as the physician seeks a mild or an active effect of ergot should he regulate his dose. In the first instance ergot is administered in doses just sufficiently large to maintain a tonic contraction of the arterioles and of the uterine tissue without producing the pain which is a constant accompaniment of the violent clonic contractions of the uterus. When the active effects of ergot are sought, large and often repeated doses are administered in such a manner as to obtain prompt and full physiologic effects of the drug.

When mild effects are sought ergot can usually be administered by the stomach. For this purpose I usually employ the purified extract called ergotin, administered in the form of capsules. Capsules containing from three to five grains each, given at intervals of four to six hours, will seldom disagree with the patient. Intervals of six hours, unless active effects are desired, are short enough. Frequently the dose of five grains will be too large for a simple tonic potion. I frequently give the ergotin in two-grain doses in capsule combined with one-fourth of a grain of the extract of nux vomica, distributing the doses so that they are taken before meals and at bedtime. Ergot can be given with good results in mild doses in rectal suppositories. Occasionally it will be tolerated in the form of the fluid extract by the mouth. There is very little occasion, however, if a patient can take a capsule, for submitting them to this nauseating dose.

When we desire to obtain the active effects of ergot some management is necessary in order to get into the system of an ordinary woman a sufficient dose of the drug, without at the same time disturbing the functions of the digestive organs.

Ergotin in capsules, in five and ten grain doses, frequently, will be tolerated by the stomach almost indefinitely. Occasionally a much smaller dose will

42

be utterly rejected. Ergotin then may be administered in eight to ten or fifteen grain rectal suppositories. The physiologic effects, in a decided manner,
may frequently be obtained in this manner. The
lower bowel should be kept clear of all fecal matter
and the suppository placed high. They may be administered as often as every six hours. Suppository
tubes may be employed to advantage for the purpose
of placing the suppository mixture higher in the
bowel than is possible with the ordinary suppository.
Small rectal enemas of the fluid extract may be employed as a means of obtaining the active effects of
the drug.

Hypodermic injections of fluid preparations of
ergot succeed in obtaining the promptest and most
efficient physiologic effects of the drug, while they
possess the objection of producing not a little pain and
occasionally abscesses. The abscesses may be avoided
by attention to aseptic principles, and the pain can be
materially avoided by selecting non-sensitive portions
of the skin combined with deep injections, and by the
employment of a mixture containing one of the less
harmful sedatives, as chloral hydrate or belladonna.

Pozzi recommends the following formula for hypodermic use:

R.   Ergotin . . . . gr. lxxv
       Chloral hydrate . . . . gr. xv
       Aqua distil . . . . ad ℥ iii
Sig.   Twelve minims injected daily.

W. H. Byford says: "Most American practitioners
now use Dr. Squibb's preparation (purified solid
extract), some of them by dissolving it in pure water,
while others add to the water a small amount of pure
glycerin. Dr. Squibb recommends a solution of this
extract as follows: Dissolve two hundred grains of
the extract in two hundred and fifty minims of water
by stirring; filter the solution through paper, and
make up to three hundred minims by washing the
residue on the filter with a little water. Each minim
of this solution represents six grains of ergot in pow-

der. Of this solution from ten to twenty minims are injected once daily or once in two days. This is the only preparation I have used in hypodermic injections, and I believe it is the best we can at present procure." It has been several years since I have used ergot hypodermically; the last that I did employ was prepared practically as given in Dr. Byford's formula.

DURATION OF TREATMENT.

As ergot at best can scarcely be termed an actual curative agent for fibroid of the uterus (except in rare instances) it follows that the duration of treatment must vary entirely with the case in hand, the results sought, and the judgment of the particular physician treating the case.

If the case is an ordinary bleeding interstitial fibroid of uniform contour, in which there seems to be no sub-mucous projections which we might hope to expell by means of heroic doses of ergot, small doses of from two to five grains each of ergotin in capsules might be given three times a day for several months. The effect sought being a general vasomotor tonic action with a special predilection for uterine vaso-constriction, and uterine shrinkage due to long continued tonic muscular contraction of the organ. The subjective results being diminution of the menstrual discharge, pressure symptoms lessened and an improvement in flesh and strength.

If the case is a sub-mucous fibroid in which the attempt is to be made to accomplish the expulsion of the mass by contraction of the uterus stimulated by heroic doses, ergotin in form of large suppositories, or better, in hypodermic injections will be administered until the result is accomplished. When the tumor is expelled the remedy is immediately suspended.

If the case is one of interstitial bleeding fibroid and the object is to control or to modify the monthly flow of blood, ergot in good full doses, either by capsules, suppository, or hypodermic injections, should be commenced a few days before the menstrual period, and

be continued until the flow has ceased, when it
can be discontinued until a week before the next
menstruation.

So that one must take into consideration the physiologic effects of the drug, under its varying doses,
take into consideration the variety and character of
the tumor and with these well in hand he must exercise his judgment in making his application in each
individual case.

RESULTS.

The results obtained in the treatment of fibroid
tumors of the uterus by ergot, depends much upon
the sincerity and peristency of the physician who is
conducting the treatment. If he is sincerely desirous
of exhausting the resources of ergot in these cases
before resorting to more radical means, or better, if he
is opposed to any more radical treatment than the
ergot treatment for fibroids, combined of course with
rational hygienic and general tonic treatment, he will
be sure to benefit a large percentage of his cases, and
possibly a small percentage will become actually cured.
Unfortunately now, in the light of the more precise
and comparatively safe surgical procedures, and the
more accurate, agreeable, if not more efficient electrical treatment, the slower, more disagreeable and painful medical treatment by ergot, is, I am afraid, too
much slighted by practitioners. As the treatment is
old, but one of the meritorious relics of pre-surgical
days, I can only indicate its real value by quoting some
of the statistics gathered by those who practiced it
enthusiastically if not almost exclusively.

One of the last statistical papers of value written on
this subject was one read at the Ninth International
Medical Congress, held at Washington in 1887, by
Prof. D. T. Nelson, of Chicago. In that paper he
reported 153 cases treated by ergot, representing the
reports of about one hundred physicians. The
method of administering the drug was seldom
described by the reporters. As a circular letter was
sent to over 4,000 physicians, and that the 100 phy-

sicians replying almost invariably reported successes, there may be grounds for believing that had many more reported the answer might not have been so uniformly favorable. The following is a brief summary of the 153 cases, as given in Dr. Nelson's words:

"The small number of cases as not affected by ergot is quite remarkable, but two of the 153 cases. All the 153 cases were benefited by the ergot, more or less, except these two. Seventy-nine were cured, tumor absorbed or expelled. In 61 other cases the tumors are smaller and their growth controlled, and there is every promise, with ergot, and perhaps without, that they will not again endanger life." There were 11 deaths in the 153 cases. "In cases 12, 13 and 17 ergot seemed to control the disease, and had it been continued favorable results were to be expected. Cases 16 and 29 died of septicemia after the expulsion of the tumor; such cases in future it is hoped, by improved methods, we may usually save. Cases 53 and 59 died only indirectly from the tumor, perhaps from embolism, the ergot having expelled or absorbed the tumor before death. In cases 79 and 123 pregnancy was an important factor in the unfavorable result." Cases 93 and 143 also died. Thus of the 153 cases, 140 remain cured or benefited.

Prof. W. H. Byford reports in his "Diseases of Women" 101 cases, including 27 of Hilderbrant's, 9 of his own, 14 of White's of Buffalo, and the remainder from a score of physicians. He summarizes them as follows:

"The total number of cases here cited is 101. Twenty-two of them are reported cured. In 39 more the tumors were diminished in size and the hemorrhage and other disagreeable symptoms removed. Nineteen of the remainder were benefited by the relief of the hemorrhages and leucorrheal discharges, while the size and other conditions of the tumors were unchanged. Out of the whole number only 21 cases entirely resisted the treatment. This shows results decidedly favorable in 80 of the 101 cases."

LECTURE V.

ELECTRICITY.

CURRENT.

In the treatment of fibroids of the uterus by electricity the direct current, or what is more commonly known as the galvanic current, is the form of electricity almost invariably employed. The maximum strength employed, 250 milliamperes or one-fourth of one ampere, requires an apparatus which will possess an electro-motive force of 30 to 40 volts. The amount of voltage required varies in different cases with the varying resistance of the electrodes and the tissues of the uterus and abdominal wall through which the current must act. This resistance may be from 60 to 300 ohms. This deviating resistance is accounted for by the use of different kinds of electrodes and the variation in the resistance of the same tissues in different individuals.

APPARATUS.

The direct current may be generated for medical uses from *a*, primary batteries, *b*, from dynamos of the non-alternating or non-interrupting variety employed for incandescent street or house lighting and *c*, storage or secondary batteries

PRIMARY BATTERY.

There are several distinct forms of primary cells employed by gynecologists in the treatment of fibroids:

Portable Battery.—The old reliable portable battery is the one with zinc and carbon elements excited by a fluid of sulphuric acid and bichromate of potassium in water in cells of glass, or better, hard rubber. The voltage of each of these cells when freshly charged is about 2 volts. Therefore a battery of this description of about 18 cells properly connected makes a very suitable portable battery for the treatment of fibroids.

There are several so-called dry-cell batteries of secret construction which appeal to one on superficial observation, because of the claims of their inventors, of cleanliness, durability, and freedom from objectionable fluids. All such batteries should be looked upon with suspicion, until they have proved themselves capable of furnishing an electro-motive force of from 30 to 40 volts for periods of five to ten minutes, several times a day for several months, otherwise the cost of recharging makes them too expensive.

Office Batteries.—For office battery where portability is not required, the Law cell, the improved Le Clanche, the Diamond Carbon, or cells of similar construction should be employed. They should be attached to a selective switchboard of such a construction that any portion of the battery may be employed at will. These cells may be placed in an adjoining closet, or cellar, and connected with the switchboard by a cable of wire, or they may be placed in a cabinet beneath the switchboard. As these cells have an average voltage of one or one and a quarter volts each, a battery of about 40 cells should be selected.

Street Wire Current.—One of the most satisfactory office fixtures for electricity is a connection from an incandescent lighting system of the uninterrupted or non-alternating variety reduced by some safe form of rheostat. One of the simplest rheostats is the McIntosh (Fig. 1). It is compact, easily comprehended and regulates the current in gradations from zero to the full strength of the street current, and reverse, without the slightest possibility of a break. A fuse box is also connected with this rheostat which will burn out and disconnect the current from the patient in case of an accidental dangerous increase of the electricity from any unlooked for source.

Storage batteries may be used as a source of elec-

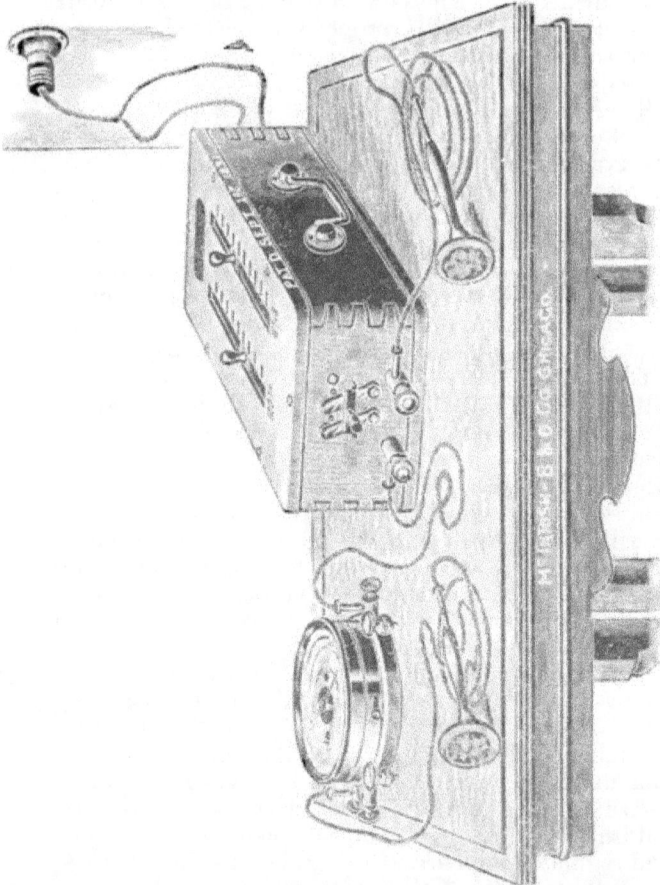

FIGURE 1.

tricity for gynecologic practice. However, as they are not economical and wherever they can be used to advantage they must be near some other source of elec-

tricity, it is obvious that one would seldom select this
form of battery.

MILLIAMPERE METER.

To employ galvanism in gynecologic practice without a milliampere meter is criminal. There are three
reasons for this: 1, because the resistance through the
abdominal walls is so small and variable, that no one.
no matter how experienced, can even approximately
estimate the amperage of a current by the number of
cells employed. 2. because of the powerful current
often required in this kind of work. a dangerous dose
might easily be given. and 3. because of inaccuracy in
recording cases.

Figure 2.

A milliampere meter should be selected which has
two readings: One scale reading to 200 milliamperes
and one reading to 50 milliamperes. This presup-
poses in reality a double instrument. By changing a
switch either reading may be selected without the
necessity of changing the connections.

The best milliampere meter I am acquainted with
is the Weston. It is reliable, convenient, double
reading, dead beat. and can be used in any position on
a level without regard to the poles of the earth. The
principal objection to this instrument for general use
is its expense (Fig. 2).

An instrument which I have employed more than
any other in my work outside of the office is the Mc-
Intosh instrument (Fig.3)   It is of the galvanometer
type and much cheaper than the Weston.   It is approximately correct, and barring the fact that it must
be carefully adjusted to the polarity of the earth, before each using, and that the indicator is not dead
beat, it is a very satisfactory instrument.   It has a
double reading; is made in two sizes, one large for
office use and one small for portable purposes.

ELECTRODES.

In describing electrodes for use in the treatment of

FIGURE 3.

fibroid tumors I will limit my description to those
which I actually employ in my own work and leave
the innumerable confusing curiosities which adorn
the ordinary instrument catalogues unmentioned.

Abdominal electrodes in use by me are of two kinds
according to the dose required.   When a current of
less than 50 milliamperes is employed a large sponge,
a large felt or a large spongio-pyoline instrument may
be employed.   These electrodes should be not less
than six by eight inches of an oval shape.   They
should be thoroughly washed in warm water before
using, and all surplus water squeezed out before the

application is made. For a current above 50 milliamperes a clay electrode or the author's membranous abdominal electrode should be employed. These being the only instruments which I have found in my experience which will uniformly distribute the current and prevent burning of the skin in spots.

The clay electrode is the cheapest form of the two. If made as recommended by Goelet, wrapped in cheese cloth with a rubber back it is comparatively clean and makes a suitable instrument where efficiency and economy alone is desirable. It is constructed of potter's clay of the consistency of putty molded into a cake, about eight by six inches in diameter by one inch in thickness.

FIGURE 4.

The membranous abdominal electrode devised by the writer, is a water electrode, the cavity of the disk holding the water being covered with animal membrane, the membrane furnishing the surface of contact (Fig 4). This instrument when filled with warm water makes an ideal electrode. It is cleanly, its temperature is easily regulated and it diffuses the current perfectly.

For internal electrodes I employ intrauterine, vaginal and rectal instruments.

The intrauterine electrodes are of two varieties; flexible concentration and soft copper.

The flexible concentration consists of platinum wire wound spirally over soft copper for varying distances tipped with hard rubber, and the portion of the instrument not active, is covered with some insulating material as rubber or linen covered with shellac. These instruments may be made of any diameter. I have them in sets of two, three and five millimeters in diameter respectively. The active surface which I ordinarily employ is four square centimeters (Fig. 5). In knowing accurately the active surface of an electrode, one can estimate more definitely the particular effect to be expected from a known current. This will be explained more comprehensively when we consider treatment technique.

Soft copper electrodes are employed in order that the uterine mucous membrane and deeper tissues may become infiltrated by cataphoresis with the salts of copper produced by a combination of tissue and cop-

FIGURE 5.

FIGURE 6.

per electrolysis, which occurs at the positive pole. I have these electrodes made in sets of three instruments, each instrument having an electrode of different caliber on either end (Fig. 6). This makes six diameters—2, 4, 6, 8, 10 and 12 millimeters. The length of each electrode surface is six inches. The portion of the staff not employed in the uterine canal is insulated with a loose rubber muff.

Vaginal electrode. – I employ an instrument for this purpose like the one shown in Fig. 7. It has an active surface of about sixteen square centimeters, the staff of it being insulated with hard rubber over a copper core. The instrument is about six inches long and about three-quarters of an inch in diameter.

Rectal electrode. – I employ a long bulbous-pointed instrument about four inches long and one-half inch in diameter for a rectal electrode. The insulated portion has a metal surface of about ten square centimeters. The staff is insulated with hard rubber.

Effects of Galvanism on Living Tissues.—Fortunately the effects of electricity upon living tissues has been so thoroughly studied, clinically and experimentally, in the last few years that we are in a position to make some pretty definite statements about its action. These studies, too, have not been limited to the living tissues, but have been carried into the chemic, physiologic and bacteriologic laboratories to such an extent that we have many experimental proofs which have proved true several former theories and exploded many others.

In applying galvanism to the tissues of the body the employment of two electrodes is necessary. In applying electricity to fibroids of the uterus the internal electrode is usually termed the active pole while the external one is called the passive pole. We speak, therefore, of three kinds of effects from the application of the current in this manner: Polar effects, inter-polar effects and general effects.

The polar effects differ materially with the pole employed. In several respects the effects at the two poles are diametrically opposed.

Effect on Sensibility.—The positive pole or anode acts as a sedative while the negative or cathode pole acts as an irritant. This effect on the sensory nerves is called the electro-tonic effect and the two effects are expressed as the anelectro-tonic effect (sedative) and the catelectro-tonic effect (irritant). The use of electricity in gynecology with its employment of large doses has abundantly demonstrated to me the electro-tonic effect of galvanism.

Effect on Blood Vessels.—The positive pole contracts blood vessels in its immediate neighborhood while the negative pole dilates them. These vaso-constrictor and vaso-dilator effects are easily demonstrated.

Chemic Reaction.—The result of tissue electrolysis

54

between the poles produces an accumulation of alkalin elements at the negative pole and acid elements at the positive pole. This results in an acid reaction obtaining at the positive pole and an alkalin reaction at the negative pole. If the electrolysis is persisted in with a powerful dose these polar accumulations become caustic acids and caustic alkalies respectively.

Effect on Tissue.—The acid accumulation at the positive pole when it becomes sufficiently concentrated from the effect of a strong dose coagulates the soft tissues and renders them for a short distance from the pole hard and dry. On the other hand the alkalin accumulation at the negative pole when strongly concentrated by a strong dose of electricity, dissolves the tissues and liquifies in the same manner as does caustic alkalies.

Effect on Pathogenic Microbes.—A zone of uterine tissue around the positive pole of a depth varying from a fraction of a millimeter to one or two millimeters according to whether the dose of current is small or great is rendered bacteriologically sterile by the employment of the galvanic current. This effect according to experiments made by Gautier, Apostoli, Enrico Burci, Vittorio Frascani and others, is not due to the electricity direct but rather to the chemic changes occurring around the positive pole as the result of electrolysis. For instance, if a copper electrode is employed oxychlorid of copper is formed as the result of a combination between the electrolyzed tissues and copper. This chemic combination is an active germ destroyer and in solution it is driven by cataphoresis into the tissues to a considerable distance, carrying its antiseptic properties with it. There is scarcely any antiseptic effect at the negative pole.

INTERPOLAR EFFECTS.

While it is easy to demonstrate polar action it is not an easy matter to make an ocular demonstration of the interpolar effects of the galvanic current on living tissues.

From experience in the employment of this current
on living healthy and pathologic tissue, experience of
many earnest investigators extending over a period
now of several years, we are convinced that evidence
enough has accumulated to justify us in saying that
the following definite effects occur in tissues so acted
upon: 1, interpolar electrolysis, 2, stimulation of
trophic nerves, 3, cataphoric action.

Interpolar Electrolysis undoubtedly occurs between the poles as well as at the metal poles themselves. When such electrolysis occurs in a fibroid
uterus it is easy to account for the reduction in size
of that growth. When the molecules of weaker ten-
acity in such tissues become decomposed into their
constituent elements ozygen, hydrogen, carbon, etc.,
these elements, as gas or solid particles, immediately
on their release become foreign substances. While
seeking for new combinations some of them are taken
into some of the many absorbents traversing the tissues and are carried out of the system. Others form
new combinations with free elements in the tissues, or
with the decomposed material of the electrodes, or
fluids surrounding the electrodes on the surface, and
still others are liberated at the poles as solids or gases.

Stimulation of Trophic Nerves. —While the electrolytic effect of the current may account for reduction or absorption of growths. I believe that this result is materially hastened by powerful stimulation of
the trophic apparatus of the uterus by electricity. We
are forced to believe this by the fact that the general
nutrition and functional activity of all the organs, any
way coming under the influence of the current, are
markedly improved.

The Cataphoric Action of the Galvanic Current.—
This is the property of a current of electricity which
enables it to push or conduct fluids in bulk through
membranous or porous conductors in the direction of
the current flow, from the positive toward the negative pole. This is also called electrical cataphoresis.
Fluids near or on the positive pole, either simple or

holding in solution drugs or chemicals, will be driven into the living tissues when living tissues are made a portion of the conductor. So that the tissues of the uterus may become impregnated with any drug which can be dissolved in water by surrounding an intra-uterine positive electrode with a film of cotton saturated with the particular fluid and causing a current to traverse the tissues.

The general effect of galvanism upon the tissues is that of a powerful tonic. Irregular practitioners, for a large number of years, employed electricity in some form successfully because of its power to stimulate general nutrition. It mattered but little what form of electricity was employed, or where it was applied so long as some portion of the sick man became a part of an electric circuit; it was sure to stimulate him, improve his nutrition and make him feel a stronger man. The powerful doses employed in the use of galvanism in the treatment of fibroids exaggerates this tonic effect of electricity to such a degree that many physicians have endeavored to attribute to it all the credit for improvement of fibroids under electricity.

Summary of Effects of Galvanism in the Treatment of Fibroids of the Uterus.

Polar Action: *Negative Pole a.* Irritant; *b.* Vaso-dilator; *c.* Alkalin; *d.* Liquifies tissues; *e.* Antiseptic (slight). *Positive Pole a.* Sedative; *b.* Vaso-constrictor; *c.* Acid; *d.* Coagulates tissues; *e.* Antiseptic (powerful).

Interpolar action: Electrolysis and trophic stimulation.

General Action: Powerful tonic.

THE APPLICATION OF GALVANISM TO THE TREATMENT OF FIBROIDS.

What is the present status of the treatment of these benign tumors by electricity? With the brilliant results of present surgery as a competitor, one must have considerable courage to offer electricity as a remedy at all in these cases. But as an abdominal

surgeon with at least average success, and at the same
time as one who interested himself early and enthusiastically in the much-lauded Apostoli treatment when
it made its *début* in this country, I am constrained by
sense of justice, knowing well both sides, to say that
in the interest of those who have fibroids of the
uterus, that the knife, even in these times of brilliant
successes in surgery, is used too often and electricity
too little. If a brilliant hysterectomy with its average mortality of 5 per cent. ended the matter, and the
95 per cent. recovering gained health immediately, we
could have but little to say. When, however, we
must reckon on the months of nervous suffering with
which the majority of these patients who have their
tumors removed have to contend, after this operation,
before they receive the well-earned cure, and when we
take into consideration the not large but certain percentage of fistulas, hernias and other well-known distressing sequelæ following operations, and last but not
least when we remember the grim specter of that 5 or
10 per cent. who did not recover, are we not justified if
we have a conscience (especially when we realize that
a fibroid of the uterus when left alone seldom proves
fatal) in giving our patients the benefit of a treatment, which seldom fails to *relieve* these cases, and
while it frequently fails to cure, *never kills* and *never
does harm* and *never interferes with the success of
an operation, if it in the end fails to cure?*

Experience in the treatment of fibroids of the
uterus by electricity has taught me how to select my
cases, when to encourage a patient to receive electricity and when to encourage her to select an operation. Rules which I have formulated and allowed to
influence me but not control me (because I make
frequent exceptions to them in individual cases) are
as follows:

WHEN ELECTRICITY IS SPECIALLY INDICATED.

1. In bleeding fibroids in women approaching the
menopause.

2. In all inoperable cases.

3. In incipient fibroids in women over 40 years of age.

4. In all bleeding fibroids of the smooth interstitial variety which have no symptoms but hemorrhage.

5. In all cases (not accompanied with pelvic pus accumulation) which refuse to have an operation.

TECHNIQUE OF TREATMENT OF TYPICAL CASES.

A typical case for the successful treatment of fibroids of the uterus by electricity is that of the interstitial variety, in which the new tissue is uniformly distributed throughout the uterus, enlarging it to a symmetrical tumor of varying sizes, and proportionately expanding the uterine canal. These cases are almost invariably of the hemorrhagic variety because of the expansion of the uterine mucous membrane. The hemorrhage occurs as an exaggerated menstrual flow. These tumors vary in size from a growth the size of one's fist to a tumor filling the abdomen with a uterine canal many inches deep. Those not exceeding six to eight inches in length and three to four inches in lateral diameter are the ones in which electricity accomplishes the best results.

METHOD OF PROCEDURE.

We seek in these cases, *a*, to transmit through these tumors, for its electrolytic effect, as strong a current of galvanism as the patient will bear, without severe discomfort, and, at the same time, not to severely cauterize the tissue at the poles. *b*, We seek to get acid accumulation at the positive pole located in the uterus, of sufficient density to coagulate the tissues and thus lessen the bleeding *c*, This same acid at the positive pole we expect to combine with the copper of the electrode and form salts, which salts in solution, by the cataphoric action of the current will be driven into the uterine tissues, immediately surrounding the electrode, and as a styptic materially aid in curing excessive flow. *d*, We seek further to obtain the powerful antiseptic effect as the result of chemic changes occurring around the internal electrode, in order to cure the endometritis which almost invariably exists as a painful accompaniment of fibroids.

After an antiseptic vaginal douche the patient to be treated is placed upon a table on her back with her buttocks drawn well to the edge and feet supported by stirrups. The size, shape and direction of the uterine canal is obtained by the use of large, flexible sounds. A large copper electrode, then, of suitable diameter, is properly shaped and passed to the bottom of the uterine canal, and the vaginal portion insulated with the rubber muff. This electrode is then attached to the positive terminal of the battery. A clay, or the writer's membranous abdominal electrode, is next passed under the loose clothing and placed on the abdomen and then attached to the negative pole of the battery.

The current is now gradually turned on while the milliampere meter is carefully watched and the features of the patient are closely scanned for signs of pain, until the current reaches 100 to 150 or even 200 milliamperes, according to the tolerance of the patient and the size of the active internal electrode. If the active electrode is of the ordinary diameter of from 3 to 5 millimeters, a current strength of 100 milliamperes can be used safely in any particular case for every two inches in length of this electrode which is active. *To be more accurate, the current should not exceed in strength 25 milliamperes for each square centimeter of active surface of the internal electrode.*

So that in the general run of cases one can safely give the patient as strong a current as she will bear without danger of producing excessive cauterization at the active pole. This will vary from 100 to 200 milliamperes. The time of each treatment should be five minutes for the maximum current employed. The treatment should be given as often as every second day. Except in cases of continuous flowing, the treatments are best given between the menstrual periods.

These cases begin to improve almost immediately. The first improvement is in relief of neuralgic and so-called pressure pain. In a few days they find that their general strength is improved. Reflex disturbances such as stomach irritation, palpitation of the heart, occipital headache and backache will be relieved. The patient will begin to eat and sleep naturally. There is a general feeling of well being engendered. In a few days the leucorrhea or purulent discharge from the endometrium will diminish. As the patient arrives near the menstrual period, she finds that the old premenstrual aches are not present, the old despondency is absent. If the treatment has been sufficiently active the menstrual flow will arrive without pain frequently. Occasionally, the first month, the flowing is fully as free as usual, although frequently it is much less. If the treatment is continued for two or three months these patients will begin to maintain that they feel perfectly well. All the old distressing symptoms will very often disappear entirely, they will gain flesh and the uterine discharge will become normal. While the tumor will still be apparent to the physician's examination it will almost invariably be found to be much diminished in size. When the time arrives in the treatment that these patients are symptomatically cured, that is when they feel no symptoms, I usually discharge them. I always inform them that the tumor has not disappeared, and that sometime it may again give them the old difficulties. As long as they are free from these they may be satisfied that the tumor is not growing—on the contrary decreasing in size. However, if the old symptoms begin to return I instruct them to seek relief again in the electricity.

The above treatment applies to the typical bleeding fibroids of interstitial variety.

Where the uterus is large and the canal is deep, it is necessary sometimes to attack the mucous membrane by piecemeal, in order to get sufficient concentration with the dose tolerated to accomplish sufficient changes in the endometrium to check hemorrhage. The concentration necessary should approximate 25 milliamperes for each square centimeter of the electrode in contact with the mucous membrane. For example, if a patient will only bear a current of 100 milliamperes, one should select an electrode of copper or zinc or platinum with a diameter of proper dimensions, insulated to all but 4 square centimeters of its distal end. The depth of the canal is measured. Then commencing with the distal end of the cavity, the exposed active surface of the electrode is made to cover in successive treatments its whole surface. By doing this the whole mucous membrane is acted upon uniformly without employing at any time a larger dose than 100 milliamperes.

INOPERABLE AND COMPLICATED CASES.

The cases which are referred to the writer for electrical treatment, in these days when active surgery offers such a large percentage of recoveries from hysterectomies, are for the most part complicated cases, which the ordinary surgeon shuns.

One complication which frequently induces the surgeon to shift the responsibility of these cases, is that of severe purulent metritis and endometritis, accompanied frequently with discharges of gangrenous masses from submucous fibroids, all accompanied with much pain, more or less hemorrhage, and with the discharges inclined to be very offensive. The patients are usually poorly nourished, with white and waxy skin in consequence of septic absorptions. When they reach this stage they are frequently pronounced malignant. The outlook for an operation certainly is not flattering.

Now the writer has been honored frequently by having such cases sent to him for electrical treatment, by different friends of his who are conscientious surgeons.

What have we to deal with? Usually a tumor of large size extending to the navel. It is soft, with nodular masses projecting from its peritoneal surfaces. The cervix is soft and patulous, with a canal large and irregular. Sometimes a small nodular mass is presenting at the cervix, This is usually soft and easily broken down. The endometrium and all cavities from which masses have been projected or from which masses have sloughed away are infected and ulcerating, and emitting a discharge which rapidly becomes offensive. From the large mucous membrane periodic and irregular uterine discharges are occurring, serving to swell the already copious outpour.

The writer has treated by electricity and symptomatically cured several of these cases in which a diagnosis of cancer had been made by men of more than ordinary talent.

I prefer, when it is practicable, to dilate the canals carefully in these cases, and remove with a dull curette the superficial débris before beginning the electricity.

I then select one of the largest copper electrodes which can be inserted and make it the active positive pole, inserting it to the bottom of the canal with its whole surface uninsulated. With the abdominal electrode in place, a current is gradually turned on until a strength of 200 milliamperes is reached, or the maximum amount under that strength that the patient will tolerate.

These treatments should be given every other day. Antiseptic douches should be employed night and morning.

These cases respond rapidly. The powerful antiseptic action on the mucous membrane makes itself apparent by the decreased odor of the discharge. The passing and withdrawing of the electrode opens and provides free drainage for the secretions. The tissues become tanned by the salts of copper which are forced into them by cataphoresis, and the discharge of blood is lessened. The patient is toned by the general effect of electricity on her system. In a word, it is frequently marvelous what a transformation will take place in these apparently hopeless cases in a few weeks of judicious galvanic treatment.

While these cases are apparently hopeless, oftentimes when they are "given over" by the surgeon, they are frequently symptomatically cured by this simple remedy. The writer has a long list of such cases, and they constitute some of the most satisfactory work he has ever had placed to his credit.

INOPERABLE TUMORS TREATED BY OTHER THAN THE INTRAUTERINE METHOD.

There is a class of complicated cases of different kinds in which it is impossible, because of the contortions of the growth, to enter the uterine canal with an electrode. Occasionally the tumor has displaced the cervix so that it is drawn high in the vagina above the bladder, out of reach of finger or sound; while again it is drawn up posteriorly with the uterine canal forming an acute angle with the vagina. In all cases where it is impossible to reach the canal, if they are treated by electricity, it is necessary to employ it without the advantages of an intrauterine electrode.

Only in the most desperate cases, in which submitting to an operation is clearly suicidal, would one think of employing electricity as a means of treatment, when an intrauterine electrode was impossible. But it is in just these cases, with their distressing neuralgic and pressure symptoms, with dyspeptic complainings and bowel irritations, the result of reflex nerve disturbances, in which an operation is discouraged, that we find patients ready to catch at any straw.

In many of these cases I believe that electricity not only offers a straw, but a veritable lifeboat to their despairing bodies.

When an intrauterine electrode is not practicable, then we should employ some other form of internal electrode which will have the effect of causing the current of galvanism to pass directly through the largest portion of the tumor.

If the vagina is not distorted so but that a vaginal electrode may be employed, that instrument should be used (Fig. 7), placing its active point posterior to the tumor. This should be made the negative pole. The abdominal electrode should be placed in such a position that the largest diameter of the tumor is interposed between it and the vaginal electrode. A current of 50 to 100 milliamperes may be safely employed. if tolerated, for a period of five minutes. The treatments may be given as often as every second day, and in a few cases every day where it is well borne.

When a vaginal electrode can not be employed to advantage in these cases, a rectal electrode (Fig. 8) should be employed. This should be placed well up opposite the tumor. It should be employed as the

FIGURE 7.

FIGURE 8.

negative pole. It should have an active surface of more than eight centimeters and the current should never exceed 200 milliamperes.

All we can expect to accomplish in this treatment is that beneficial action derived from passing a strong direct current through any tissue containing muscles, nerves. lymphatics and blood vessels, viz., a powerful trophic stimulation to the part, and incidentally a powerful general tonic effect on the general system.

These cases get great relief. Neuralgias stop. Troublesome abdominal reflexes cease. Circulation is improved. Nutrition is stimulated. Sleeplessness disappears. Bowels are stimulated and relieved of troublesome distension symptoms. The tumors often seem to decrease in size. The degree to which each

of these symptoms are relieved varies, of course, much
in individual cases. The writer has seen a large number of cases completely and for an indefinite time,
relieved of all these symptoms. In fact, some of the
most gratifying cases of relief he has, are of this
variety. Their cases are apparently so hopeless that
often any relief is very gratifying.

LECTURE VI.

SURGICAL ENVIRONMENTS; OPERATING ROOM; STERIL-
IZERS; STERILIZING INSTRUMENTS, LIGATURES AND
HANDS; CATGUT PREPARATIONS; PREPARATORY
AND AFTER TREATMENT OF PATIENTS.
DRAINAGE.

The environments of a patient who is about to submit
to a surgical operation for a fibroid of the uterus must
be made surgically clean. These environments include
operating room, bed, sterilizers, instruments, ligatures
and operators' and assistants' hands and clothing.

OPERATING ROOM.

In a private house a room should be selected which
has direct light through one or two large windows; a
room which can be stripped of furniture, hangings
and carpets. It should be convenient to the bedroom
of the patient, or better the bed can be placed in the
room in readiness for use when the operation is finished—the operating room constituting the bedroom.
The woodwork of this room should be thoroughly
scrubbed with soap and water, and the walls and ceil-
ing carefully wiped free of dust. The room should
be thoroughly aired by opening the windows and a
reliable means of heating should be at hand in order
to render it dry and to keep it at a temperature of
80 degrees F. when required. The table, which is
selected for the operating table, and the stands for
instruments and dressings, together with all recep-
tacles or slop tubs and basins should be carefully
scrubbed and then conscientiously wiped with a 1:500
solution of chlorid of mercury. All tin, iron or porce-
lain basins should be boiled for one-half hour in a wash

boiler or other large boiler, as a means of sterilization

The bed, if possible, should consist of a hair mattress which has recently been purified by steam. In a hospital a large steam sterilizer should be provided where hair mattresses can be sterilized frequently. The bed should be completed with dry sterilized sheets, blankets and pillow slips. If there is no sterilizer at hand the bedding can be sterilized by boiling in water one-half hour, and drying in a pure room, and ironing with a hot iron by an intelligent attendant or nurse. Gowns, towels and aprons should be sterilized in the same manner as the bedding, provided there is no regular steam sterilizer at hand.

In an institution the operating room should have floor and walls of such material that they can be thoroughly washed with antiseptic solutions and provided with a central drain which will allow the cleaning of the walls and floors with water direct from a hydrant through a hose. The drain should be reliably trapped, or better, drain directly in to the external air. For convenience, a perfectly fitted operating room should have several anterooms, including a preparatory room where the solutions are prepared, the water sterilized, and where the heating apparatus for the sterilizers and the sterilizers themselves are located. This room should have washable walls. There should also be one or more anesthetizing rooms, and finally there should be convenient dressing and wash rooms for the surgeon and his assistants. The private operating room which I use at the Woman's Hospital is shown in Fig. 9. It has direct side light and a large skylight. Its walls and floors are of marble. It is lighted at night entirely by incandescent electric lights, gas being impracticable where an anesthetic is necessary; these lights are in abundance, so that an operation can be performed equally well at night or day. The preparatory room is adjacent. This is shown in Fig. 10. It is entirely in marble. The battery of Boeckmann's sterilizers is shown in the foreground. In the farther end are two large tanks in which the water is sterilized for the operation, one being filled with cold sterilized water and the other with hot water. They are connected with the operating room by large faucets which pass through the wall. Directly off from this room is an anesthetizing room, and adjacent to this are two dressing rooms with washing utensils. In the operating room is a spectators' rail which separates the operator, assistants, nurses and all operating paraphernalia from those who may be invited to witness operations.

Sterilizers.—In a private house, in emergency cases, an ordinary copper or tin wash boiler may take the place of the most elaborate sterilizer. The gowns, towels, gauze operating sheets and all large articles used externally can be thoroughly sterilized by boiling for thirty minutes. For sterilizing instruments, silkworm gut, silk and other smaller articles a smaller kitchen article such as a sauce pan, or porcelain-lined flat pan, may be utilized as a sterilizer.

In large institutions large steam sterilizers are employed. I have used the Arnold sterilizer for dressings and instruments and other small articles until quite recently, since which time I have adopted for my hospital work the Boeckmann steam sterilizer (Fig. 11). These sterilizers are simple in construction, durable, inexpensive, efficient for all work, even the sterilization of catgut, and they possess the advantage of sterilizing with steam, while at the same time when the process is finished the articles are left perfectly dry.

At the Woman's Hospital several of these sterilizers are employed and everything that is liable to be required in several operations is sterilized, and the unopened sterilizers are placed in the operating room for future use. Surgeon's and nurses' gowns, towels, gauze, silk and silkworm gut in cotton-stoppered test tubes are removed as they are required at the time of the operation, while a separate hot water sterilizer (Fig. 12) is employed immediately before the operation for sterilizing the instruments.

Ligatures.—I employ braided silk, silkworm gut

Figure 9.

FIGURE 10.

and catgut for sutures and ligatures. Silk and silkworm gut I sterilize by boiling, or by steam in the
Boeckmann sterilizer. They are placed in small skeins
in large test tubes loosely stoppered with cotton and
subjected to a temperature of boiling water for twenty
minutes on two successive days when I have no sterilizer
at hand, and to the temperature of the superheated
steam for a like length of time in my hospital work. I
only open at the operation a sufficient number of tubes

FIGURE 11.

for the operation in hand, the balance being reserved
for future cases. Tubes of silk and silkworm gut may
be prepared in considerable numbers and sterilized by
steam with an efficient cotton filter and afterward carried to operations anywhere. On opening, the cotton
stopper is first burned down low with the tube, then
removed and the skein of material carefully lifted out

with sterilized forceps and placed in sterilized water when it is ready for use.

Catgut is the form of absorbable ligature which I, employ for buried sutures. I have it sterilized in the Boeckmann sterilizer with dry heat at a temperature of 284 degrees F. for a period of three hours. Previous to sterilization, the catgut, cut in suitable lengths, is wrapped in oiled paper, one thread in each paper and the paper enclosed in small hermetically sealed envelopes. While this accomplishes perfect sterilization, as can be demonstrated by bacteriologic tests it has been argued that sterilized catgut may act as a very favorable nidus for the growth of pathogenic germs in tissues in which it is buried, tissues which

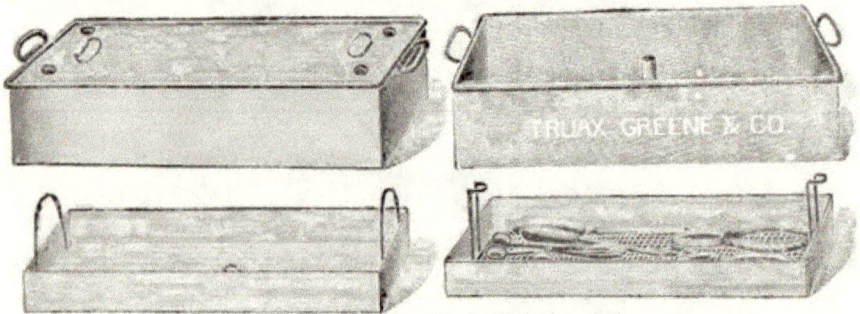

FIGURE 12.—Boeckmann's Instrument Sterilizer.

without the presence of this perfect sterilized culture medium would be competent to resist the few germs left in a wound, after ordinary surgical precautions had been exerted. For this reason I not only render my catgut aseptic with heat but I supplement that process by saturating it with non-poisonous antiseptics. According to Arthur Woodward Booth's admirable article in the *Therapeutic Gazette*, December, 1894, he found that pyoctanin blue in a 1 to 1000 alcoholic solution will render catgut thoroughly antiseptic and at the same time impart to it a longer life. Pyoctanin is a much more powerful antiseptic than chromic acid and therefore may be employed in more diluted form. Compared with bichlorid of mercury it is a more perfect germicide, non-poisonous, and it imparts a longer life to the gut. Catgut saturated with pyoktanin becomes an antiseptic suture, the antiseptic of which can in no way prove a source of danger.

THE WRITER'S METHOD OF CATGUT PREPARATIONS.

A skein of new catgut is cut into about four lengths. This makes the threads of about forty inches each. Each section of the skein is twisted into a loose knot and they are soaked in ether for twenty-four hours in order to remove the fat. It is then boiled in alcohol, in a closed jar, for one hour in the steam sterilizer. Before boiling in alcohol the bunches are divided into their separate threads and each thread twisted into a little coil. After the sterilization by alcohol it is carefully removed from the jar by an intelligent conscientious nurse, with sterilized forceps in sterilized hands, to a jar containing a solution of pyoctanin 1 to 1000 in absolute alcohol. Here it is allowed to remain for twenty-four hours in order to become thoroughly saturated with that powerful antiseptic. I then have it distributed into small wide-mouthed one-half ounce bottles, containing oil of juniper. Into each of these bottles about four of the forty-inch strands are placed. The bottle is then corked with a rubber stopper and is not opened until it is to be used at an operation, when the catgut is threaded directly from the bottle. The bottle, or several bottles, with their rubber stoppers may be immersed in a 1 to 1000 bichlorid solution on the instrument table ready to be opened by the surgical nurse in the course of the operation. After it has once been opened I discard any small amount of catgut which may remain after the operation, preferring to use always from a fresh supply.

Instead of boiling the catgut in alcohol it may be sterilized in closed envelopes in the Boeckmann sterilizer as before described, at a temperature of 284 degrees F., and then treated from that point in the same way as that sterilized by boiling in alcohol.

It seems to me that this is an ideal and simple method of catgut preparation. There can be no doubt of its absolute sterilization after it has been boiled in alcohol for two hours, or after it has been submitted for three hours to a temperature of 284 degrees F. in the sealed envelopes in the Boeckmann sterilizer.

Dr. Booth found that pyoktanin permeated every fiber of catgut when it had lain in a 1 to 1000 alcoholic solution of the drug for twenty-four hours.

Dr. Booth quotes Sternberg as quoting Jaenicke on the antiseptic properties of pyoctanin as follows:

Staphylococcus pyogenes aureus restrained by solution of 1 to 2,000,000; bacillus anthrax aureus restrained by solution of 1 to 1,000,000; streptococcus pyogenes aureus restrained by solution of 1 to 333,000. In blood serum stronger solutions are required.

Thus we have not only a sterile catgut, but we have that sterile catgut thoroughly saturated with an efficient and non-poisonous antiseptic. The juniper oil preserves the catgut indefinitely, it fixes the pyctanin so that it will not stain the hands, and it keeps the catgut soft and pliable.

The life of the catgut in the tissues prepared by this method is a little less than that prepared by chromic acid, but considerably longer than that prepared by bichlorid of mercury. By placing it in small bottles it can be handled economically without the necessity of ever being obliged to open twice the same supply.

Sponges.— Sea sponges are not safe when prepared under the most careful supervision, whereas gauze sponges may be perfectly reliable whenever there is the simplest device at hand in which they may be boiled.

The best gauze sponges are made from loose-mesh gauze folded into three or four thicknesses, with the edges fastened with a running stitch of cotton thread. They may be made of any size. These sponges are sterilized in a steam sterilizer, or in emergency cases they may be sterilized by boiling with the instruments. In laparotomy cases I prefer those sterilized by the

A MPHITHEATER OF POST-GRADUATE HOSPITAL.

dry method. They are then used but once and are discarded. This is not economical because of the large number of sponges frequently required, nor is it necessary. The advantage possessed by the dry sponge is in the increased absorptive power of the gauze.

Gauze.—The writer has devised an apparatus in which to sterilize gauze for operations, either for hospital operations or operations away from home. Chas. Truax, Greene & Co. kindly constructed this apparatus for me, as well as the furniture shown in the operating room in Fig. 9. It consists of a little stand which fits into the catgut sterilizer of the Boeckmann apparatus, or which can be set into any steam sterilizer (Fig. 13).

FIGURE 13.    FIGURE 14.

The stand contains seven large test tubes, two inches in diameter, and about eight inches in length. In each of these tubes can be placed all the gauze of any one kind that will be required at any ordinary operation, which is about two yards of sheet iodoform gauze. Ordinarily I have two tubes filled with iodoform sheet gauze, one with plain sterilized gauze, one with one-inch strip iodoform gauze cut the strong way of the cloth, one with two-inch strip iodoform gauze cut in the same way, and one with a skein of silkworm gut and a skein of braided silk. These tubes are loosely filled and their mouths closed with cotton. They are then subjected to steam sterilization in the Boeck-

PRIVATE OPERATING ROOM, POST-GRADUATE HOSPITAL.

mann or other steam sterilizer at maximum heat for one hour. They are then set aside and the following day they are again subjected to the superheated steam for one hour, and then dried by removing the cork in the top of the Beckmann sterilizer so as to get the action of the dry heat. The contents of the tube are now thoroughly and permanently sterilized, and will remain so for weeks if the cotton stoppers in the mouth of the tubes are not removed.

When I wish to preserve these tubes for indefinite use I have the nurse slip a sterilized rubber cap over the cotton and the end of the tube before removing them from the sterilizer. They may then be set aside for an indefinite time.

When I wish to operate away from the hospital, I place the rack containing the tubes required into a metal box (Fig. 14), and that in turn is stored away in my instrument bag.

Preparation of Operator and Assistants. — It should not only be taken for granted but should be insisted on, that any and all persons participating in the high calling of surgery should take a general bath, including the hair, every day.

Dress. For important operating, such as we have to deal with, special dress for the operator and assistants is indispensable. Suits of white ducking or linen should displace the street apparel. Over this sterilized gowns should be worn. In this dress the operator can be comfortable and do hard work in a temperature of 80 degrees F. When he is through operating all wet clothing, made so by perspiration and the fluids of the operating room, can be replaced by his ordinary dry out-door dress and the reminders of the operating room are left behind.

Preparation of Hands. — After the nails are filed short and smooth the hands and forearms should be thoroughly scrubbed for fifteen minutes in hot water with a stiff nail brush and plenty of pure soap. The water should be changed at least five times. The time should be estimated by an actual time piece and not by guess work. A nurse should supervise this part of the work in important operations, and report to the responsible chief any laxity on the part of any participant. The spaces beneath the nails should be thoroughly brushed and the undersurface of the nail scraped with a steel nail cleaner. After the soap and water scrubbing, the hands should be washed in alcohol and then immersed in 1:1000 bichlorid of mercury solution, and this solution brought into contact with all irregularities by means of the hand brush. The hands should, finally, be rinsed in warm sterilized water. Before beginning the operation the hands should be rinsed in hot water which is placed in a basin close to the operator, so that it may be used from time to time during the operation. After the hands are once washed they should not be allowed to come in contact with anything before or during the operation which is not surgically sterile.

Arrangement of Operating Room.—The steam or dry heat sterilizers containing dressings should be convenient to the nurse. Reservoirs of sterilized water, hot and cold, should be placed near the sponge table. Two large glass irrigators should be at hand. The table with which the Trendelenburg position may be obtained is necessary, and should be placed in an advantageous position for light and assistants. For a laparotomy, the arrangement of the furniture and participants of the operating room should be approximately as follows: The table near the center of the room with the head of the patient near the chief window. Anesthetizer at head of patient. Operator on right of table (from head). Chief assistant and assistants opposite the operator with the chief nearer the head of the table. The surgical nurse in charge of instruments at stand to right of operator. Nurse in charge at foot of table with sponge dish on small stand in reach of second assistant. Assistant nurse to her right, the latter to work sponges, and to attend to irrigators, sterilized water, etc. Superintending nurse without regular assignment, ready for emergency. To left of operator, small table with sterilized solution for hands. Back of the assistants a similar table. Visiting physicians, admitted after everything is ready for the operation to begin, are arranged around the room out of reach of the operating corps, or any concerned in the operation.

If the case is one where a vaginal operation is required, the head of the patient is directed away from the window, and the patient in the exaggerated lithotomy position is placed with the buttocks directed toward the light. The limbs are supported on either side by two assistants. The operator sits at the foot of the table with the instruments at his right hand. To the left is the nurse with sponges and the irrigator.

PREPARATORY TREATMENT OF PATIENT FOR LAPAROTOMY OR VAGINAL OPERATIONS.

Kidneys. — The failure to recognize obscure kidney disease in patients before submitting them to a severe operation has been the cause of many avoidable deaths. We should not only recognize kidney difficulties in every case but we should also know when a case is laboring under some form of kidney trouble, whether that stage has been reached beyond which it is safe to proceed. It is not enough that the urine in any given case is approximately of normal quantity, of approximately normal specific gravity, and that it gives negative results in tests for albumin and sugar. It is necessary to learn the history of the case, to estimate the specific gravity in a twenty-four hour specimen, to ascertain the amount of urea for twenty-four hours, and supplement this with a thorough and complete microscopic examination.

In diabetes we should not operate. In interstitial nephritis when the disease is not far advanced an operation may be risked with proper preparatory treatment. These latter cases are the very ones which from their great difficulty of diagnosis are often neglected, and consequently disaster results. The importance of the subject must be my excuse for entering into

primary details. The following summarizes the signs
of chronic interstitial nephritis: Lowered specific
gravity of urine; patient arising at night to void
urine (when there is no bladder or urethral disease
to give rise to such a procedure); an enlarged heart
with accentuated second sound; a tense pulse and
diminished urea. Albumin is frequently absent. The
diagnosis is doubly sure when hyaline casts are
found.

Every patient should be scrutinized in all these
points. If the foregoing state of affairs exist to a
marked degree I refuse to operate. If, however, with
the above symptoms I find a normal quantity of urine,
which does not show a reduced specific gravity under
1010 to 1014 and the amount of urea does not sink
lower than six or seven grains to the ounce, if the
patient is well preserved generally without advanced
heart disease, I am confident that I can operate with
safety, if I can secure proper preparation.

I prepare these patients, first by placing them on
an exclusive farinaceous diet with milk and fruit for
an indefinite number of days before the operation. A
week or ten days before the operation a diuretic is
added with instructions to drink large quantities of
water, the object being to increase the daily quantity of urine from 60 to 100 ounces, in order to thoroughly flush the kidneys and rid the patient of dangerous accumulations. With 60 to 100 ounces of
urine flowing for several days, with the patient living
on a non-nitrogenous diet, with the urea in improved
proportion considering the diet, I feel safe in risking
an operation.

Dr. Charles W. Purdy, who has had an enormous
experience in watching the behavior of kidney diseases
under operations, says in reference to *chronic paren-
chymatous* nephritis: "I see no reason why these
cases, if unaccompanied with dropsy may not be operated upon if carefully selected."

Bowels.—In preparing patients for an ordinary
laparotomy I begin preparations of the bowels two

nights before the morning of the operation. The
first point is to seek thorough emptying of the bowels
throughout their entire length. The second point
should be to render their contents thoroughly aseptic
and the third should be to impart to them a maximum
tonicity.

The bowels are emptied by means of mercurials and
salines. The first night of preparation, six grains of
blue mass are given. The next morning at 6 A.M., one
drachm doses of citrate of magnesia are given every
hour until the bowels move, or feel as though they
would move with the aid of a small enema. This
ought to insure a thorough movement of the entire
length of the intestinal canal. If the movements are
such, with the above treatment, to insure a thorough
evacuation, and to start a free flow of bile, as indicated by the yellow glistening appearance of the stool,
no farther catharsis is necessary. The lower bowel
should be thoroughly evacuated, however, by the
employment of large enemas of soap and water,
repeated four or five times during this second day of
preparation. The last enema should be given late in
the afternoon of this second day of preparation, if the
operation is to be done the following morning, and
the next morning if the operation is to be performed
in the afternoon. The bowels are rendered aseptic by
large doses of bismuth and salol. During the first
and second days of preparation, gr. x of salol and gr.
xx of subnitrate of bismuth should be given every
six hours.

The bowels are stimulated by means of carminatives,
alcoholic stimulants and strychnin. The second day of
preparation 1 drachm doses of tr. of cardomon in one
ounce of brandy are given every six hours. Strychnin
is commenced three days before the operation in 1-40
gr. doses every eight hours, and gradually increased in
quantity until 1-20 gr. doses are given. The bowels
should be kept in a thoroughly aseptic condition by
feeding the patient a milk diet for two days before
the operation.

External Preparations of the Patient. —The first day of the preparation the patient should receive a thorough general bath and then be placed in clean clothing and a clean bed. The abdomen should then be rubbed with a saturated solution of permanganate of potassium until it is of a uniform mahogany color. This should be scrubbed off by means of a sponge or brush and the application of a saturated solution of oxalic acid. A green soap compress should be bound on the abdomen, this latter to remain all night. Vaginal douches of, first, soap and water; second, 1:5000 bichlorid solution, and third, plain sterilized water should be employed this first night. These should be repeated the night before the operation, and a last vaginal douche given immediately before the operation. The second night of preparation should begin with shaving of the abdomen and pubis, and should be followed by applying a bichlorid compress. Immediately before the operation, after this compress is removed, the abdomen should be scrubbed with green soap and hot water, this to be followed with alcohol or ether, and covered with an antiseptic towel until the incision is made.

The bladder should be evacuated by means of a catheter immediately before the operation.

Dress.—The patient is to be put in a long, loose, woolen night gown immediately after an operation, and is thoroughly covered, except the abdomen, with flannel blankets during the operation.

Operation.—I am suspicious of an operator who operates on time. The best operators are those who operate well in the smallest space of time; this implies that the best operators are slow operators. An abdominal incision should be a clean, true, unhaggled cut, so that accurate coaptation is possible. Cold sponges should be employed on the external incision in order to contract the capillary vessels and check their bleeding without the necessity of forceps. Forceps should be employed, however, everywhere in abdominal surgery that their use will save blood, because most of our old-fashioned shock was caused by unnecessary loss of blood. Keep the operative field free from oozing points if possible, even at the loss of a little time. The peritoneum is best opened between two catch forceps elevated so as to present a thin fold. After a small opening is made, the finger passed into the cavity should act as a guide upon which to complete the incision. The peritoneal edges should be attached to the integumentary edges by means of catch forceps, to prevent its peeling off from the abdominal walls in the subsequent manipulation. In all pelvic surgery of smaller tumors the pelvis should at this point be elevated by means of the Trendelenburg table; the elevation being sufficient to draw the bowels away from the field of operation, and to elevate the contents of the pelvis. Sterilized silk or catgut should be employed for any pedicle which it is safe to tie and drop. Catgut may be employed to close simple peritoneal rents.

Drainage should be employed in all cases where extensive enucleation has occurred, where there is a slow venous oozing from separated adhesions or where aseptic matter has in any way contaminated the peritoneum. Drainage, in competent hands, never does any harm, therefore, where there is the slightest doubt, it should be employed. It has saved many lives, and made more comfortable those who might not have died without it, but who have been given the advantage of it.

After my operation is finished, the peritoneal cavity is thoroughly dried; then if there has been at the operation a process of enucleation, leaving of necessity slight oozing points, or in cases where ordinary adhesions have been separated, after drying the cavity as far as possible, I place in the cul-de-sac a glass drainage tube and pump out any remaining fluid. I next protect the abdominal contents from the abdominal wound with a large flat sponge, and insert the sutures after which I again pump the drainage tube. If there is more than a drachm of bloody fluid,

I leave the tube in until the sutures are nearly all tied and the sponge removed; then I make a last trial of the tube. If the fluid amounts to one-half drachm or more, and is bloody, I allow the tube to remain; if, on the contrary, it is nearly dry, or the contents is simply colored water, the result of flushing, if it has been employed, I remove the tube. What has been done influences one in regard to drainage. I almost invariably drain after it has been necessary to flush. I believe the peritoneum is satisfied, to an extent by the flushing, and will consequently neglect, in a degree, to absorb any remaining fluid. Experience seems to sustain that argument. Mikulicz drain is almost indispensable in a limited number of cases. Cavities may be packed with gauze which can not be reached with glass drainage tubes. Hemorrhages in cavities so packed will cease, when a glass drain would not avail. Operations are now possible with the Mikulicz drain which were impossible without it. The question about drainage is not, shall we drain, but how, and how often.

To Prevent Intestinal Obstruction. In abdominal surgery one is constantly watching the behavior of the intestines. They are our prominent point of attack in our preparatory treatment, they are our greatest source of anxiety during the operation, and upon their management after the operation much watchfulness is imposed. All of this anxiety is caused by our desire (with the exception of care against wounding when operating) to prevent obstruction. It, therefore, is a point in the technique of this work to which discussion may profitably be directed.

The pathology of obstructions is well summed up by a valuable contribution on this subject by Dr. Ashton, of Philadelphia, from which I quote:

"Adhesions between the intestines and raw surfaces: *a*, to an omental stump; *b*, to the edges of the vaginal wound following supra-pubic or vaginal hysterectomy; *c*, to a pedicle; *d*, to raw surfaces on the intestinal wall.

"2. Paralysis of the intestines.

"3. Local spasm of the intestines.

"4. Impacted feces.

"5. Bands of inflammatory lymph.

"6. Adhesions between coils of intestines or between the gut and neighboring parts, due to traumatic inflammation.

"7. Kinking or twisting of the intestines, due to faulty technique.

"8. Including the intestines within the loop of a suture of the abdominal wall, or between the edges of the abdominal incision.

"9. Slipping of a coil of intestines through a slit or an aperture."

Under the first head, "Adhesions between the intestines and raw surfaces," we must seek our remedy during and following the operation. Intestines should be handled and exposed as little as possible in order not to produce hyperemia or denudation of their surfaces. An omental stump of any considerable size should be selveged by inverting its raw edges with a running catgut, or with ligature. When denudations of the pelvic or intestinal peritoneum can not be reinforced by a superabundance in the neighborhood, care should be taken to carefully arrange the intestines in as near the normal position as possible. A pedicle of large size should be covered by securing over its end the peritoneal covering with a running stitch of catgut. Raw surfaces of any considerable size on the intestines should be covered with peritoneum if possible, with the edges well secured. Paralysis of the intestines may be avoided by emptying them thoroughly previous to the operation of all irritating matter (which may ferment and cause distension) by rendering the contents aseptic by means of bismuth and salol, and the employment of full doses of strychnia to act as a muscular tonic. Carminatives, such as wintergreen, cardamon, etc. may also be employed as antiseptics and muscular tonics. During the operation the intestines should not be handled or chilled in order to avoid paralysis. After the operation, nourishment of non-fermentive and easily absorbable nature should be employed. The bowels should be stimulated to early action in order to keep them empty and avoid the beginning of distention, which soon leads to paralysis. An early movement of the bowels, or free passage of flatus, assures a normal disposition of the bowels as regards location. If they adhere after such time, it will be in an advantageous, not cramped position.

A flat sponge beneath the abdominal wound, after carefully spreading down the omentum and before the wound sutures are inserted, will avoid including an intestine within the loop of a suture, or between the edges of the abdominal wound.

When ventral fixation of the uterus is practiced, great care should be exercised in disposing of the intestines in such a manner as to avoid their slipping through the opening left between the uterus and the abdominal wound.

AFTER-TREATMENT.

The immediate after-treatment consists in stimulating the patient out of any tendency to nervous shock which may exist. She should be surrounded in bed with dry heat, and in hospitals placed on a water bed. If there has been any considerable loss of blood, the feet may be elevated in order to restore blood pressure in the brain. In severe cases of shock from loss of blood, it is well to bandage the blood out of the lower extremities by means of elastic bandages. A saline solution under the integument may assist to fill the blood vessels. Oftentimes the difficulty is not lack of fluids so much as lack of tone, which allows a patient to bleed to death, as some one has put it, "into her own dilated capillaries and venules." Here direct arterial stimulants and vaso-constrictor remedies are called for, as well as strong nerve stimulants. In these cases, I immediately order hypodermics of nitroglycerin, strychnin and digatalin. Stimulating enemas of whisky and warm water may also be given What is done for shock should be done promptly, as patients who are allowed to go on for a few hours with a sub-normal temperature and high pulse, are with great difficulty restored.

Dressing Glass Drainage Tube.—The glass drainage tube, when it is allowed to remain, should be emptied with a syringe with a long rubber nozzle the first time in one hour. If the fluid is more than a drachm it should be dressed again in an hour, if a drachm or less, the interval between dressings should be increased one hour, and the same rule followed until the fluid is less than a drachm and of a light amber color, and the interval from four to six hours. At this time the tube may be removed. If it is left longer than thirty-six hours, a piece of sterilized gauze should be put in its place for six hours, when the latter is removed and the wound is closed with slight pressure and its closure is obtained by extra pressure of external straps.

Care of Capillary Gauze Drain.—If capillary gauze drain has been employed instead of the glass drainage, the protruding gauze (from vagina or abdominal wound) should be kept abundantly covered with a pad of loose fluffy gauze, and this should be changed as often as it becomes saturated with fluids. If all drainage ceases in twelve to twenty-four hours as indicated by dry dressings, the gauze packing may be removed. However, if drainage is free and the patient is normal it may remain forty-eight to sixty-two hours. When it is possible this drainage should have its exit through the vagina. After it has been removed a loose gauze packing should be placed over the wound.

Dressings.—The wound is closed with silkworm gut, including all parts of the wound edges, and dusted with sterilized iodoform, and the dressing is a thick one of iodoform gauze, held in place by adhesive straps tight enough to take the strain off the sutures without puckering the integument. Over this is placed a liberal allowance of absorbent cotton,

and all retained with a binder. The dressings are not
disturbed for four days unless there is pain or temperature. At the end of the fourth day the dressings
are carefully removed, and the wound is thoroughly
but carefully washed with equal parts of 95 per cent
alcohol and 1:5000 bichlorid solution. Iodoform is
again applied and the dressings renewed. The seventh
day, before the stitches are removed, I again have the
wound washed in the same manner and, after their
removal, dressed as before.

In vaginal operations the vaginal wound is dusted
with sterilized iodoform and the vagina loosely packed
with strip gause. It is removed in forty-eight hours
and after twelve hours vaginal douches of bichlorid
of mercury solution followed by plain water are employed once or twice daily.

Bowels — If flatus has not passed freely, per rectum,
in twelve hours by the simple employment of a rectal
tube, I employ the "one, one, one" enema, one ounce
of sulphate of magnesia, one ounce glycerin and one
ounce water. If this does not start the gas in two
hours, I order it repeated in double quantity. If this
enema is not retained, and flatus has not passed, I
order an enema of soap and water one pint, with one-
half drachm of turpentine. If they are still obdurate,
I begin one-half grain doses of calomel in ten grains
of bicarbonate of soda, given every two hours for four
doses, or until gas passes, alternated with drachm
doses each of gran. citrate magnesia and sulphate
magnesia in an ounce of water. Following these remedies in one hour, another "one, one, one" enema is
given. It must be an obstinate case indeed that will
not yield under the above remedies. If the stomach
is irritable and will not tolerate the bicarbonate of
soda, the calomel may be given dry on the tongue.
Other salines may be substituted for the above if
they are objectionable. The bowels should be moved
from above on the fifth day with small doses of
some effervescing salt, or, if required, a more vigorous
laxative.

Diet. First week, fluids; second week, semi-solids; third week, semi-solid and solid food in small quantities; fourth week, good substantial food, with a few curtailments. As soon as the patient is out of the anesthetic, I begin to give hot water in teaspoonful doses as often as every fifteen minutes. If the stomach tolerates this, the quantity is increased to one-half ounce, and the interval may be increased in length. If the patient is nauseated and the hot water causes vomiting (and it should be hot water), or increases the nausea, it should be withheld. When the patient can take the hot water, and still complains of thirst and begs for cold water the nurse is instructed to let her rinse her mouth with cold water. Ginger ale is a good alternate with water for the first twenty-four hours. After twelve hours, drachm doses of peptonized milk may be sandwiched with the water. If peptonized milk is offensive, plain sterilized milk may be substituted, or sterilized milk and lime water. Milk in some form, I feel to be the most perfect food. It should be increased every hour until $\frac{1}{2}$ ounce doses are given by twenty-four hours, to one ounce by forty-eight hours, and to two ounces by the end of sixty-eight hours. Barley water may be alternated with the milk. Later, the monotony may be relieved with the meat and shell-fish broths, thin gruels, etc. The fourth or fifth day the patient may be allowed to extract the juice of broiled beef by chewing it; the fiber, of course, should be rejected. Tea may be given the second or third day as a relish. Orange juice, and the juice of other fruits, may be given in small quantities the third or fourth day. Rules can not be laid down in regard to the diet of these patients, general principles only can be hinted at. The patient should be seen each day and her wants studied. If stimulants are required, one of the best is good brandy. Champagne is all right if its sweetness does not make it objectionable. If patients are unable to retain enough by stomach to properly nourish them, enemas of either milk or stimulants should be resorted to.

Getting Up.—Uncomplicated laparotomy cases are gradually bolstered up until they can sit in a bed with a bed rest at about the fifteenth or sixteenth day, sit up in a large chair at the twenty-first day, and leave the hospital from the twenty-eighth to the fortieth day.

LECTURE VII.

THE AUTHOR'S OPERATION OF VAGINAL LIGATION OF THE
BROAD LIGAMENT AND OTHER MINOR OPERATIONS.

Vaginal ligation of the contents of the base of the
broad ligaments, for the cure of fibroids of the uterus,
was devised and performed by me as a new and original
operation Nov. 15, 1892, and was described and published in the April number of the *American Journal
of Obstetrics* in 1893. In the January number of the
American Journal of Obstetrics, 1894, I reported six
cases treated by the new operation.

The operation as originally described by me is as
follows: The ligation of more or less of the broad
ligament of the uterus, with its vessels and nerves,
the extent of the ligation depending upon the result
sought, from a simple ligation of the base of the ligament, including the uterine arteries and branches of
both sides without opening the peritoneum to a complete ligation of the ligament of one side, including
both uterine and ovarian arteries, with partial ligation
of the opposite ligament without opening the peritoneal cavity, if possible, but by doing so if necessary.

The results sought in the operation are, first to
check uterine hemorrhages by cutting off blood chan-
nels, and secondly to produce atrophy of the fibroid by,
1, depriving it of nourishment through the blood vessels and, 2, by changing the nutrition of the uterus
by interfering with its nerve supply.

Immediately after publishing my first article on
this operation there were two claimants for priority;
Dr. Walter B. Dorsett, of St. Louis, and Prof. S.

Gottschalk, of Berlin, Germany. Dr. Dorsett, in a
letter to the *American Journal of Obstetrics*, claimed
that he had suggested a similar procedure to my operation in an article he published in the St. Louis *Cour-
ier of Medicine* in 1890, the article bearing title of
" A Case of Atrophy of the Female Genitalia following Pregnancy and remarks." In this article he
made the following observation: "I believe that in
the treatment of uterine fibroid : . . to ligate
the uterine artery would not be an unscientific procedure. On the contrary the more I have thought
of it the more I am inclined to believe that it would
be the most certain mode of treatment." Dr. Dorsett, while advancing the theory, had not at that
time carried it out on a living woman.

Prof. Gottschalk based his claim of priority on an
article read by him at the Brussels Congress, Sept.
16, 1892, with the following title: · Die Histogenese
und Aetiologie der Uterusmyome." In the latter
paragraphs of this article he casually suggested ligation of the uterine arteries and stated that he had
performed the operation twice. This is what he said:
" The bilateral ligation of the uterine arteries appears to be the therapeutic measure in this regard
for the earliest incipient stages of myoma. This
offers no difficulties in its technique; it is easily
performed in a few minutes. . . I have already
performed this ligation in two cases in which I was
able to early diagnose the development of multiple
myoma with best results."

Thus these two men both suggested tying the
uterine arteries for the cure of fibroids and at least
one of them (Gottschalk) performed the operation
twice before I described my operation. This would
definitely decide the question of priority in their
favor if the operation they suggested was identical
with mine. Their operation is not identical in
theory, in execution, or in description with mine.
and therefore their claim of priority for my operation can not be substantiated.

The operation suggested by these men simply includes the ligating of the uterine artery from the vagina, while, 1, I ligate in all cases, the whole base of the broad ligament, in order, *a*, to occlude not only the main channel of the uterine artery, but all collateral branches; *b*, in order to destroy the function of the nerves as well as the arteries of nutrition; *c*, in order to diminish nerve reflexes. 2. I include, in desperate cases, not only the base of the broad ligament with the uterine artery and branches in my ligatures, but when practicable ligate high enough on one side to take in the ovarian artery. 3. I advise accomplishing this result, if possible, without opening the peritoneal cavity, but by doing so, if necessary.

TECHNIQUE OF OPERATION.

The preparation of a patient for vaginal ligation of the broad ligaments of the uterus should be similar to that demanded for vaginal hysterectomy, as described in my Lectures VI and IX. Ether is used as an anesthetic and the patient is placed on the operating table in the exaggerated lithotomy position with buttocks brought to the end of the table, with an assistant on either side to support the limbs and hold the vaginal retractors. A broad, short vaginal retractor above and below exposes the cervix, which is transfixed with a strong silk ligature to be employed in handling the uterus. The uterine canal is dilated and the uterine cavity curetted with a dull curette and thoroughly irrigated with 1:1000 bichlorid solution and then loosely packed with iodoform gauze. This procedure cleans the uterus and makes it impossible for the vaginal wounds to become infected by a septic uterine discharge. The uterus is now drawn down in order to put the broad ligaments on the stretch and then drawn to the right side so as to expose the left vaginal vault. The mucous membrane of the vagina at the utero-vaginal fold on the left side is then caught with a tenaculum and incised with a pair of

curved scissors. One blade is allowed to enter beneath
the mucous membrane and a curved incision one and
one-half to two inches long is made over the broad
ligament and at right angles to it (Fig. 15). By
means of the index fingers of the two hands the oper-

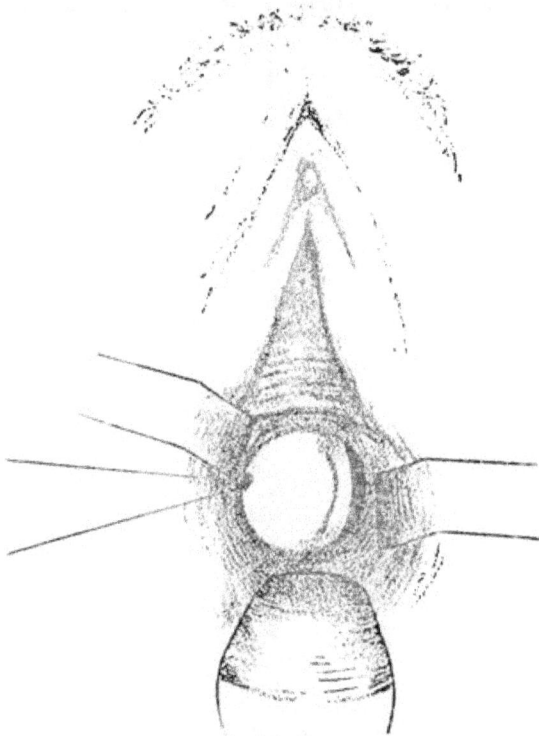

FIGURE 15.

ator now separates the vaginal tissue from the broad
ligament and carefully separates the broad ligament
in front from the bladder for a height of two inches
and laterally for nearly the same distance (Fig. 16).
The bladder should be carefully separated in this way
in order to avoid the danger of wounding the organ,

96

and by pushing the separation laterally the ureter is forced out of danger. One then carefully separates the broad ligament posteriorly to the same height as in front, without, if possible, penetrating the peritoneum. Now, by passing one finger behind the other in front, the whole base of the broad ligament, representing (two-

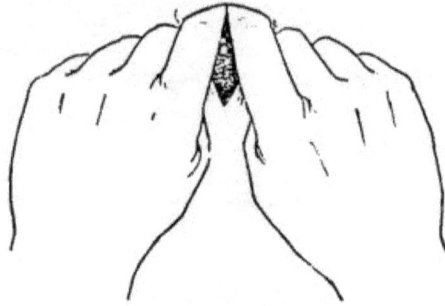

FIGURE 16.

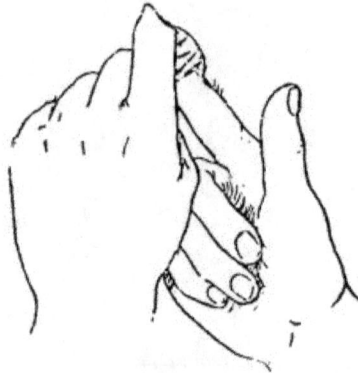

FIGURE 17.

thirds of its bulk, can be grasped (Fig. 17) for a distance of an inch to an inch and a half from the uterus. In this grasp one can easily feel the throb of the main trunk of the uterine artery and occasionally several branches. The curved pedicle needle is then passed, armed with No. 10 silk, strong pyoktaninized catgut or kangaroo tendon, and guided by the index finger of the left hand (Fig. 18) is made to penetrate through the broad ligament. The ligature is drawn through, the needle removed and the base of the broad ligament is thoroughly ligated at a distance of one inch or more from the uterus. The ligature is cut short, leaving it

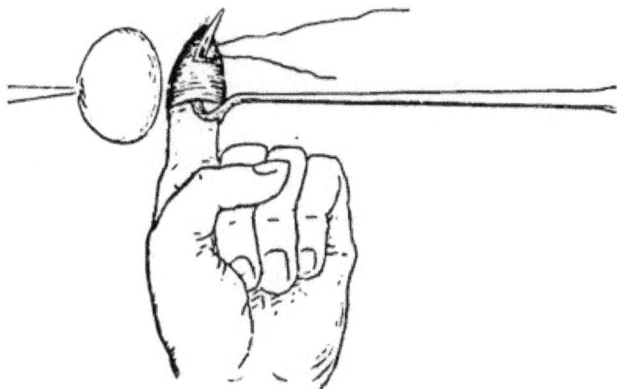

FIGURE 18.

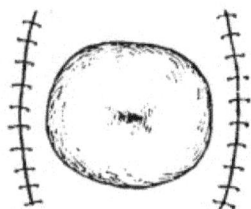

FIGURE 19.

buried in the tissues. The other broad ligament is treated in the same manner; the vagina is well sterilized with bichlorid solution and the vaginal incision accurately approximated with fine antiseptic catgut so as to completely bury the broad ligament ligatures (Fig. 19). The handling string is now removed from the cervix, and the end of the gauze strip packed in

the uterus is tied to another strip and the vagina is
filled loosely with a gauze drain.

The after treatment of these cases is very simple
The vaginal and uterine gauze is removed the second
or third day, and twice a day thereafter a bichlorid
vaginal douche 1:2000 followed by plain douche are
given. Figure 20 shows the position of ligatures
when only the base of the broad ligament is ligated.

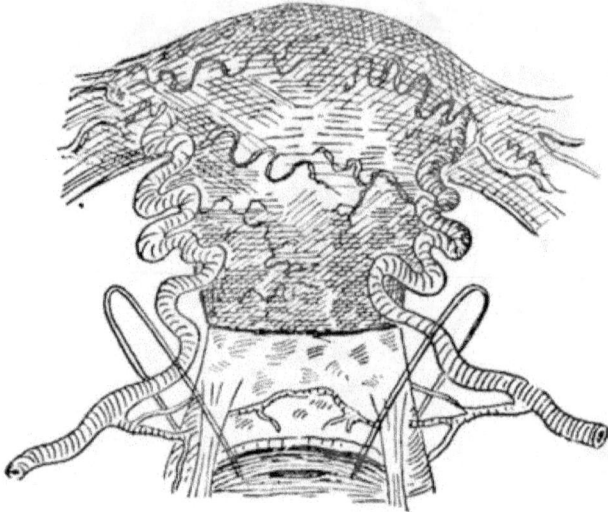

FIGURE 20.

SELECTION OF CASES.

Interstitial fibroids of the uterus of moderate size
are the cases in which the best results will be obtained
by this operation. Subperitoneal fibroids springing
from the fundus of the uterus especially would
scarcely be benefited to any great extent by depriving
the lower part of the uterus of its nourishment.
Neither would one expect to obtain any lasting benefit from this operation in cases of pedunculated submucous fibroids. On the other hand, in true intersti-

tial growths depending upon the whole uterus for their nourishment, cases where the tumor is the uterus, and these represent 75 per cent. of all fibroids of the uterus, wherever it is possible to tie the base of the broad ligament from the vagina, this operation may be expected to accomplish prompt and decided relief of symptoms and a rapid reduction of the tumor. The cases in which the most satisfactory results must be expected are incipient or small fibroids of the interstitial variety which show themselves late in the menstrual life. Here, we have a uterus which is small enough so that it has not risen above the brim of the pelvis, one which can be easily reached from the vagina so that its broad ligaments are accessible from below. Such a fibroid, too, from the age of the patient will reach a state of quiescence as soon as the menopause is established. In such cases, then, a major operation is particularly undesirable, because it is not imperatively demanded and because of a reasonable chance of relief at the approaching change; on the other hand the symptoms (with severe hemorrhage usually as the principal one) are such that immediate relief is earnestly sought, if one can be reasonably certain of obtaining it without submitting to a dangerous and radical procedure. These are ideal cases for this operation.

Another class of cases in which this operation has been employed with gratifying success and in which it will probably find favor with the most radical operators, are those of continuous and profuse hemorrhage in which the desperateness of the drain is such that the patients are depleted to such a degree, that no radical procedure can be thought of, until a minor operative procedure has checked blood waste and recuperation is accomplished. My fourth and sixth cases were like the above. In the fourth case hemorrhage was very profuse and the patient was completely exsanguinated and so weak that she had not been out of bed for several months. Some time before I determined to submit her to my operation an attempt had been made to remove the appendages, or. if possible, when the laparotomy was in progress, the uterus. From complicated adhesions and the weakness of the subject neither operation was possible after the abdomen had been opened. The tumor was large and the elevation of the uterus in consequence was great, and it was with the utmost difficulty with the aid of the most competent assistants, that I finally succeeded in ligating thoroughly the base of each broad ligament. Both ligaments contained several arteries, some of them as large as the normal radial artery. They were all tied in mass. Hemorrhage stopped from the instant of tying the last ligature and it has never recurred. It has now been over three years since I operated on this case. The uterus has reduced until it is but slightly larger than normal. The woman (I examined her but a few months ago) is perfectly well. She has a slight menstrual flow each month, and is free from pain.

Case six was of a severe hemorrhagic nature in a typical interstitial fibroid of three by five inches in diameter. The woman was too weak and depleted for a radical operation. I did my operation on her and the result was marvelous. In three months' time she had recuperated so that any radical operation might have been done without danger.

Dr. Humiston, of Cleveland, reported to me a case in which he used my operation as a procedure of last resort, in a patient nearly moribund from hemorrhage. She was so weak that he only attempted ligation on one side. The woman stopped bleeding instantly and eventually recovered. Hence, the operation may with propriety be employed as a rational temporary expedient in desperate cases of whatever variety, where uterine blood loss is conspicuous.

CASES.

In selecting cases for this operation I have been very careful. In the majority of them I have operated on, there seemed no alternative. All were des-

perate ones, like cases 1, 2, 3, 6 and 8, or they would
not submit to a more radical procedure, and milder
means, as electricity, ergot, etc., would not accomplish
satisfactory results. I have been more conservative
in adopting the operation, I am afraid, than the results
in the few cases I have operated on would justify.
One reason for not adopting the operation in a larger
number of cases is that I wished first to learn of the
remote results. It is now over three years since my
first operation and most of the operations which I
have performed were during the first year. I have,
therefore, a three years' history to analyze in the
majority of my cases. In the following report there
are no instances in which, at least, a year has not
elapsed since the operation.

Case 1. This was an interstitial fibroid in a maiden lady 36
years old, in size extending above the umbilicus. The hemorrhage was exhaustive and the patient greatly reduced in
consequence. Her heart was hypertrophied and her condition was such that no surgeon with a proper care
for his statistics or his patient's life would have ventured a hysterectomy. She was operated on by my operation Nov. 15, 1892. The hemorrhage decreased about one-
half for several months after the operation. The tumor in
the first four months materially decreased in size. In May,
1894, the hemorrhage is reported much modified, and no
longer a source of alarm. The patient at that date considered
her condition greatly improved, hemorrhage cured, tumor
materially reduced and pressure symptoms subsided. March
12, 1896, four years and three months after the operation the
patient reports herself well. The original fullness produced
by the tumor she can no longer feel. No pain. The last flowing of any consequence was November, 1894. Since then the
flow has been very slight until last July, when it practically
ceased. "I have color in my lips and cheeks. I walk two
miles or more every day," she writes. This report is certainly
very gratifying.

Case 2. The second case was a married woman 40 years of
age who had been under electrical treatment for a hemorrhagic
myofibroma of the uterus. The galvanism decreased the size
of the growth but did not materially lessen the exhaustive
hemorrhage. The tumor was of the interstitial variety and
the uterus appeared the size of a three months' pregnant
uterus. When the patient entered the Woman's Hospital for
operation December, 1892, she had been having almost continuous hemorrhage for several months. Upon exposing the

uterus with the retractors at the time of the operation, the cervix was large, blue and vascular. As the vagina was large the operation was very easily executed. The ligature on the left side included fully two inches in width of the broad ligament at a distance of at least an inch from the uterus. When I tightened this first ligature one of the spectators, a well known gynecologist, called my attention to the fact that the cervix had perceptibly paled in appearance. The broad ligament was easily exposed on the right side, and fully as much of it ligated as on the left. If there had been any doubt of the procedure affecting the vascularity of the uterus, it vanished when the second ligature was tied. The cervix immediately paled until it was nearly as white as a piece of cartilage.

The covering of the broad ligament was so loosely attached in this case that I could easily feel the main channel of the ovarian artery, and it would have been an easy matter to have included it in the ligature.

After over three years I can promise this case a perfect cure. The uterus has reduced to normal size. The hemorrhage has ceased completely. All pain has disappeared. A slight menstruation, normal in quantity, occurs each month. The patient's health has improved so, that from a state of almost complete invalidism she is transformed into to a strong healthy woman. The improvement has been progressive from the day of the operation. I have seen this case within the month (March 1896).

Case 3. This patient was operated on in January, 1893. She had an incipient interstitial fibroid of two years' standing which was profusely hemorrhagic in nature. I tied the base of both broad ligaments including the uterine arteries and their branches. The relief was immediate. The menstruation for the next four months was scanty. The patient gained in health and strength rapidly. The tumor, which was the size of a four months' pregnancy at the time of the operation decreased markedly in size within three months. Four months after the operation I lost track of this case, as she lived in a distant State and neglected to keep me posted. Her last letter gave a report of perfect health.

Case 4. This patient had a large, bleeding fibroid filling the pelvis, which extended to the umbilicus. The uterus and appendages were firmly adherent and immovable. Laparotomy had been attempted on the case, with the object of removing the appendages or the tumor. The abdomen was opened, but the adhesions and unusual complications rendered it impossible to remove the tumor or even accomplish the oblation of the appendages. The patient was so unusually reduced from loss

of blood at the time of my operation that she had not been
able to be out of bed for three months.

I operated on the patient in January, 1893, at the Post-
Graduate Medical School of Chicago. The operation was
accomplished with great difficulty because of the large size and
immovability of the uterus. Finally, however, after consum-
ing more than an hour in time I succeeded in ligating thoroughly the two broad ligaments well above the uterine arteries
and their branches.

In June following I made the following report on the case :
"The flowing ceased immediately and the patient was relieved
of her drain for over two weeks. She then had a few days'
flowing, which resembled an ordinary menstruation. She has
rapidly and steadily improved since that time. She has menstruated regularly but scantily, and without pain. She can at
this time (June, 1893), five months after the operation, attend
to her duties as a housewife, and considers herself cured. The
tumor has become reduced in size until it is no longer noticeable as a deformity, and so that the patient herself is no longer
conscious of its presence."

Since the foregoing report was written in June, I
have seen this patient several times, the last time
within the month. The patient was then examined
by several physicians, one or two of whom on inde-
pendent examinations, failed to notice any abnormal
enlargement. The uterus is still somewhat larger
than normal, but is not more than three or four inches
in diameter, while the testimony of at least three experienced diagnosticians will bear me out in the estimate
that its former diameters were not less than four and
a half by eight inches. The patient is in good health
now, Jan. 1, 1895; menstruation is regular but scanty,
and she is free from pain. The patient, so far as I
know, has remained well.

Case 5. This case was a woman with an interstitial fibroid
about three by five inches. She was about 30 years of age, and
the growth had been noticed for three years. Her principle
symptoms were profuse menorrhagia with severe menstrual
pain. The case was referred to me by Dr. F. H. Greer, of Colum-
bus, Neb. I did my operation on the woman Jan. 8, 1893. She
had a little subsequent temperature, and one month after the
operation the ligature sloughed from the left broad ligament.
Four months after the operation Dr. Greer reports the woman
well. "Menstruation scanty, no pain. Fibroid diminished in
size until the uterus is about normal. Patient claims that she

is cured." This report was made in June, 1893. I have been
unable to get any history subsequent to that date.

Case 6. - This was the wife of a physician of more than ordinary ability and reputation. The patient was about 36 years of
age, slightly above the average height, with well-proportioned
frame, but poor in flesh, with a skin blanched and a body almost
exsanguinated. The uterus was about the size of a three
months' gravid uterus. The tumor was uniform and evidently
interstitial. The uterus was in normal position. The cervix
was nearly two inches in diameter, the os patulous.

The history of the growth dated back, undoubtedly, several
years. The patient had borne no children. The menstruation
had for nearly two years increased in quantity and duration,
until now, while coming with absolute regularity, it lasted fifteen
days, and that in spite of vaginal and uterine tampons, the recumbent position, ergot, hydrastis and the rest. She flowed each
month until she was completely exhausted, scarcely recovering in
the next thirteen days sufficiently so that she could assume the
upright position without fainting. Accompanying this unusual
discharge was uterine pain, which in its severity brought the
patient to the point of unconsciousness. During the four days
in which the woman could drag herself around in the latter
part of each intermenstrual period she did so with the greatest
discomfort on account of the pressure and neuralgic pains of
the pelvis. Upon examination of the broad ligament from the
vagina the finger could detect on either side the large, pulsat-
ing artery as it fed the tumor. The latter was movable, the
appendages apparently normal, the broad ligaments accessible.
In fine, here was an ideally typical case—a hemorrhagic fibroid
of the uterus, a bed-ridden patient, an authentic diagnosis, an
unusually interested physician to carefully watch and estimate
the result, and one who enthusiastically demanded a trial of the
new operation. Under the circumstances it seemed to me that
much depended upon this case, as though the fate of this operation must necessarily be more than usually linked with this
particular patient.

I operated on this case Aug. 2, 1894, at the Chicago Hospital,
with Dr. Robert Dodds and Dr. Okischet as assistants. The
left broad ligament was carefully dissected from the peritoneal
covering behind, and from the bladder in front, until fully
two thirds of it could be grasped by placing one finger behind
it and another finger or instrument in front of it. When
grasped in this manner several beating branches of the uterine
artery, together with the main artery itself could be detected.
This entire mass was then ligated in two sections with No. 12
braided silk, the silk cut short, the parts irrigated and the
vaginal wound closed with catgut. After treating the opposite
side in the same manner, the vagina was cleansed and loosely
packed with iodoform gauze. When the operation was finished
the throbbing arteries, which could be distinctly felt before,

could no longer be found. The cervix, which was large and purple previous to the operation, became pale and cartilaginous in appearance as soon as the ligaments were secured.

The patient remained in the hospital three weeks. The first menstruation was due the day following the operation. It began the next morning, but was so slight and painless that the patient would not believe that it was her menstruation until several days had elapsed and no other flow appeared. It lasted about three days and was barely perceptible; absolutely no pain. The after treatment consisted in vaginal douches after removing the gauze, light diet and the recumbent position for two weeks.

August 30, the second menstruation reappeared; there was a little of the old pain, but not sufficient to require anodyne of any kind; the flow was half the usual amount and lasted six days. September 28, the third menstruation appeared; the amount was normal in quantity, lasting but four days; the pain was slight. October 26, the fourth menstruation appeared; the amount normal in quantity, lasting but four days; the pain was slight. The patient was seen and examined by me just before the last menstruation. She had gained several pounds in flesh, her cheeks and lips were red and she was a picture of health and robustness. Her feelings were in accord with her appearance, as she enthusiastically assured me that she felt perfectly well. On examination I found the uterus was reduced in size. It was little, if any, larger than normal. Its bulk had decreased one-half. The cervix was small and normal. No arterial pulsation could be felt in either broad ligament or around the vault of the vagina.

The next report I received was in January, 1894: " I have to report," the husband says, " that Mrs. X. menstruated from December 19 to 24. That the amount was about the same as before, i. e., slightly above the normal. Pain rather excessive for two days (possibly due to rheumatism and neuralgia). After flow had ceased I examined and found ligature in

vagina and also small sinous opening to left side of
cervix. Since then there has been slight discharge
from same. She had been suffering some pain at
that point, no pain since ligature came away." He
adds enthusiastically: "Taken all in all, the result
so far is a grand success." Jan. 17, 1894, he writes:
" Mrs. X. is up to-day (the fifth day) after the easiest
menstruation she has had in her life; pain moderate
and only one day. This in face of the right side still
discharging. In the next two months I expect to
have a well woman. The uterus is now practically
normal."

I have lost track of this patient entirely, and I
regret that I am unable to complete so interesting a
history. If the husband of this patient should read
this report I hope that he will communicate with me.

Case 7. Mrs. S., Denver, Colo., aged 35, uterus about double
the normal proportions, containing two or more centers of development and an extremely hemorrhagic tendency, was the
seventh case operated on. The case had been treated unsuc-
cessfully by curettement, electricity and the ordinary remedies
for checking uterine hemorrhages. The uterus was retroverted
but free from adhesions. The patient was prepared carefully,
and at the Woman's Hospital, on Nov. 11, 1893, I ligated the
base of both broad ligaments, and shortened the round ligaments. The uterus was drawn well down, and each broad
ligament, after incising the mucous membrane covering them
in the vault of the vagina was dissected free from the bladder
and rectal attachments and then ligated with two strong ligatures. These ligatures were placed high enough to include the
uterine artery, all its branches, and all of the contents of the
base of each broad ligament. The ligatures were cut short
after they were tied, the mucous membrane of the vagina was
reunited with a running catgut ligature and the vagina packed
with iodoform gauze. The round ligaments were then shortened and the uterus left in a position of anteversion. Three
days later the gauze was removed from the vagina, an antiseptic douche was given and a Smith-Hodge pessary was inserted. The antiseptic douches were then continued daily.
The first menstruation was due four days after the operation.
It did not appear. The second menstruation also failed to
appear, notwithstanding the fact that menstruation had ordinarily been exhaustive.

One of the wounds caused in the operation for
shortening the round ligament suppurated, and obliged

the patient to remain in the hospital until the latter part of January. Dizziness was complained of about the time when the menstruation was due. This symptom continued with different degrees of severity for some time, gradually disappearing. February 13, three months after the operation, the first flow appeared. The patient writes: "First menstruation came on the 13th of this month, without pain, but quite profuse for first two days. Since then has continued, including to-day (the 18th). Discharge light."

March 19, 1894, the patient reports: "Am feeling fairly well this month. Had pain in back with last menstruation, which commenced March 13. First three days quite profuse; last four days very little No dizziness this month." April 18, she writes: "Menstruation came on four days in advance of schedule time; continued one week. Am in fairly good health."

December 14, 1895, two years after the operation, the husband writes that his wife suffers considerably with vertigo, especially severe immediately before menstruation. "The operation performed by you has in a measure been successful, as the menstrual discharge is much less than before the operation and the womb is in much better position."

March 26, 1895, the husband writes: "Her menstruation is not profuse and she has less pain; her general health about the same" (as in her last letter). "I think the riding of the bicycle improves her general health and strengthens her in those parts wherein she is weak."

Case 8. Mrs. Z., Muscatine, Iowa. About 35 years of age. No children. Multiple fibroid of the uterus approximating in size a four months' pregnancy. Hemorrhage profuse, followed for a week by excruciating pain. Patient became extremely exsanguinated at each menstrual period. Frequently the flowing would last for two weeks. The uterus had been curetted. Electricity failed to control the hemorrhage and only partially modified the pain The irregularity of the uterine canal undoubtedly accounted for the failure of the electricity. Nov. 28, 1893, the patient submitted to my operation for ligation of the broad ligament. The tumor was developed more to the

left side into the left broad ligament. I succeeded in separating
the broad ligament for a height of two inches. On the right side
a large double ligature was employed, while on the left side first
a double and finally a second one higher and farther away from
the uterus was applied. The ligatures were cut short, the vaginal vault closed with catgut and the vagina packed with iodoform drain. The first menstruation was due three days following the operation. A slight watery discharge occurred instead
of blood. Two days following the operation the patient complained of pain similar to that which ordinarily occurred
after menstruation. Feb. 1, 1894, the patient's husband writes:
"She commenced her menstruation January 25, and it was
continued until to-day, February 1, one day less than last time.
Had one day of some pain; not bad. She is getting stronger
and can get around the house without being very tired, although
she has not yet ventured out." February 26, the report is:
"Mrs. Z. was sick this time six days, the same as last time
Had considerable pain two days which was very severe, the
same as she complained of before the operation. She is getting along very nicely. She is now able to go out, and takes a
walk every day." March 26, the husband writes: "I am
ready to make another report, but not as good a one as I would
like. Mrs. Z. was sick on time and the flow was very little
compared to what it has been, lasting but three days, but she
had a great deal of pain some before she was sick, and it was
quite bad for two days after the menstruation. . . . Everything seems to be working very well if she could only get rid
of that pain." April, menstruation still decreasing in quantity; the pain decreasing. "There was one day of pain," the
husband writes, "and the flow amounted to but very little."
May 6, he writes: "Mrs. Z. has been feeling splendidly all
this last month. Last week was her time to be sick again.
The flow did not amount to anything, just enough to show.
. . . In regard to her general health, it is excellent. Eats
well, sleeps well and goes out every day the same as other
women. Has gained her natural amount of flesh and a little
more." I examined the patient May 19. The uterus was reduced in size one-half. Patient in perfect health.

March 13, 1896, two years and four months after
this woman's operation, I received the following
report from her: "Since the operation I have gained
twenty pounds or a little more up to date. My menstruation period is about one-half the time and amount
it was before the operation. The pain is very much
less than I had before the operation, but it has not
left me altogether. . . . I have the strength of
the average woman now, while before the operation I

was compelled to be in bed over half the time. Between my menstruations I enjoy as good health as any one could ask. . . ."

Case 9. Mrs. C., aged 41, a resident of Iowa, consulted me for a bleeding painful fibroid of the uterus in May, 1894. The tumor was interstitial, uniform in contour, enlarging the uterus to the size of a four months' pregnancy. The hemorrhage at menstruation was profuse and lasted six or eight days at a time. The menstrual periods were accompanied with considerable uterine contraction pains. The patient complained of a great deal of heaviness in the pelvis and pains caused by the pressure of the tumor. The patient was weak, rather exsanguinated and nervous. I concluded that the case was a suitable one for my operation. The operation was done May 19, 1894. The lower portion of the uterus was so large and filled the pelvis so completely that it was with a great deal of difficulty that I accomplished the satisfactory ligation of both broad ligaments. However, when the operation was finished I was well satisfied that both uterine arteries had been thoroughly shut off.

The patient improved from the first. There have been no more hemorrhages. I have examined the patient two or three times since the operation, once within six months. The tumor has decreased in size, but has not disappeared. The pains and pressure symptoms are much better. The woman is apparently a healthy woman and does very much as other healthy women. In reply to my letter of inquiry she said March 12, 1896, one year and six months after her operation: "Since the operation I have had but few hemorrhages, while previous to that I had them very frequently. I am now quite regular, though I never go to my full time—about three weeks. I have less pain, but the heaviness still remains. I am better in health and strength than before the operation."

Judging from my other cases I expect this woman to gradually recover. My fear was, when I adopted this operation, that collateral circulation would speedily overcome the result of ligating of the blood supply. Experience, however, shows that the results of the operation are greater the farther away from the operation we get.

Case 10.— Mrs. B., aged 35, a resident of the central portion of the State, came to the Woman s Hospital in August. 1894, to consult me about an interstitial fibroid. She had borne no children. The uterus was large, regular in contour, hard and about four by six inches in diameter. It was freely movable in the pelvis. The woman gave a history of severe monthly hemorrhages which lasted anywhere from six days to two weeks at a time. Accompanying the flooding were severe contraction pains. The woman was bloodless, pale, weak and extremely nervous. In all other respects she was normal

I did my operation on the case in August, 1894, with the assistance of the house staff of the Woman's Hospital. It was easily performed on account of the movability of the tumor and the looseness of the broad ligaments.

September, 1894, the patient wrote: "It is now six weeks since I have menstruated. The pains are not any better My bladder trouble (pressure) is much relieved." October, 1894; "I have menstruated since my last letter. The quantity and length of time was small. Had a good deal of pain the first two days."

November 13, 1894. "Menstruations three weeks apart. I flowed more than usual."

December 17, 1894. "Pains some less. My changes came at the correct date, but was greater in quantity than it should have been"

January 29, 1895. "I flow a great deal more than I think I ought. I have to change my napkins six or seven times a day."

March 25, 1895. "I was a little better my sick week this month. The flowing did not last so long as it did before my operation, but more than is right. My pains are gradually improving."

May 27, 1895. "I flowed very freely and had a great deal of pain this month."

Case II. Mrs. Y., the wife of a very intelligent physician of Indiana, consulted me in November, 1894. She had an interstitial fibroid about the size of a four months' pregnancy. She was 43 years of age. I operated on her Nov. 7, 1894. Both broad ligaments were tied, so as to include two-thirds of their bulk. This occluded the uterine arteries on both sides with all their anomalous branches. I have seen this patient several times since her operation and the uterus is gradually lessening in size and the patient's symptoms are subsiding. I expect this case to prove successful with a little more time. In reply to a request from me for a statement of progress, the husband writes March 12, 1894, a year and four months after the operation, as follows: "My Dear Doctor: In reply to yours of yesterday, I have to say that my wife, on whom you operated Nov. 7, 1894, is doing very well as far as the fibroid is concerned. It has decreased in size some, not a great deal. The menstrual flow on two occasions was quite profuse, but the last two periods have been very scanty only lasting three days, and only using one or two napkins in a day, whereas before the operation she used eight and ten each day for four or five days. She suffers very little pain, in fact none for the last month. Before her operation she suffered constantly. Her general health has greatly improved, and she has gained ten pounds in flesh, is much more cheerful, and in fact improved in every way. . .

My wife is now past 45 years of age, and I believe if the tumor does not increase in size until after the menopause, she will entirely recover."

Case 12. - Miss V., single, age 40, consulted me in November, 1894, on account of a painful bleeding fibroid. She was depleted to an unusual state, and her nervous system was a wreck. She had an interstitial fibroid with the canal of the uterus measuring four inches in depth. The uterus measured approximately three by six inches in diameter. The organ was movable. Hemorrhage occurred only at regular menstruation periods. At this time it lasted a week or ten days and was very profuse. Accompanying the flow was great prostration of the patient and also most excruciating pelvic pressure symptoms. This condition of affairs had been going on for months until the patient from blood drain and harassment of pain had been brought to a deplorable state of health.

December 5, 1894, I ligated the base of both broad ligaments. The operation was accomplished with ease because the uterus was movable and the broad ligaments loose. The patient recovered nicely from the operation. The next two or three menstruations were much more normal, the quantity of flow being very small and the pain scarcely perceptible. The woman was placed on tonics and urged in every way to increase her blood supply. Her nervous system reacted slowly. Her menstruations later became more profuse and was accompanied on several occasions by quite severe pain. While but a short time has elapsed since the operation the patient is gradually improving.

March 12, 1896, one year and three months after the operation, she writes: "The amount of menstrual flow averages about one-half the amount it was before the operation. Have gained a little in flesh. Have considerable pain. Still have nerves although under better control than formerly."

This patient has improved in many ways. During her intra-menstrual periods she is comparatively well and is able to go about and to do more work than she should attempt, whereas previous to her operation she was unable to do much. While she had been neglected for a long time and her health had reached a low ebb, I can not but believe that she will gradually improve as a direct result of the diminished flow. Her tumor decreased in the first four weeks fully one-third. It has not increased perceptibly to the patient since she left my care.

Case 13. Miss B., age 26, consulted me in January, 1895, for a bleeding intramural fibroid. The uterus was about five inches long and had a canal three and three-quarter inches in depth. The canal was a little irregular. In the fundus of the uterus could be felt two distinct centers of development, one on the anterior surface about two and one-half inches in diameter, and projecting from the main body of the uterus one and one-half inches. It was hard and had the unmistakable firm consistency of a fibroid mass. On the posterior surface of the fundus at its junction with the neck was a second center projecting from the uterus about one and a half inches. This mass was irregular and was fully two inches in diameter. The symptoms which brought this patient to me were prolonged and exhaustive hemorrhages and uterine pains. The lady is a vocalist of unusual talent and these symptoms interfered seriously with her profession. The case was a typical one for hysterectomy, especially as the left ovary was enlarged and cystic, but as that would involve the removal of the ovaries the patient objected to this because of the popular but unfounded fear that removal of the ovaries impairs the voice. I therefore decided to perform my operation on the case. When the patient was under the anesthetic I confirmed absolutely by bimanual manipulation my diagnosis as given above. February 1, 1895. I operated on this patient. She left the hospital in two weeks. She did not have an unfavorable symptom. Menstruation practically ceased from the date of the operation. There was but the slightest show each month. No pain whatever. In less than a month she was able to attend to her professional duties and was stronger than she had ever been. This perfect condition of affairs continued until about Dec. 8, 1895, ten months after her operation. At this time I was called because of a sudden attack of severe pain she had experienced in the left side of the pelvic region. The pain was accompanied with profound prostration and shock. I diagnosed ruptured cyst of the left side and advised a laparotomy. In making my examination I was surprised to find a perfectly normal uterus. I performed laparotomy on this patient Dec. 28, 1895, and removed a ruptured ovarian cyst of the left side and punctured a small cyst in the ovary of the right side. This gave me an opportunity to examine the uterus which I had treated by my operation a little over ten months before. On the anterior portion of the organ corresponding to the location of the anterior fibroid described above I found buried in the wall and projecting a half inch, a fibroid center one-half inch in diameter. On the posterior surface, corresponding to the other center which

I palpated at the previous operation, was another center distinct but even smaller than the anterior one. These were both exhibited to the house staff and physicians present at the operation. The behavior of this case was most gratifying until the complication of the ruptured cyst arose. This fortunately gave me an opportunity of examining by direct sight the results accomplished by the first operation.

I have no doubt but that those two fibroid centers would have been starved out eventually and the case actually cured without any further interference.

General Summary: Thirteen cases operated on in which more than a year has elapsed since the operation:

Case 1. Age 40. Operation Nov. 15, 1892. Very large bleeding fibroid. Present condition: Tumor much reduced. Hemorrhages ceased. Patient well.

Case 2. - Age 40. Operation December, 1892. Fibroid interstitial, size of three months' pregnancy. Profusely hemorrhagic. Present condition: Tumor disappeared. Absolute cure.

Case 3. Operation January, 1893. Interstitial bleeding fibroid of two years standing. Four months after operation. Last report: Tumor reduced, patient much improved.

Case 4. Age 38. Operation January, 1893. Very large adherent interstitial fibroid. Excessively hemorrhagic. Patient bed-ridden. Two years afterward: Uterus reduced almost to normal size. Hemorrhage ceased. Patient well and strong.

Case 5. Age 30. Operation Jan. 8, 1893. Interstitial fibroid three by five inches in diameter. Profuse hemorrhage. Report four months after operation: Uterus normal: hemorrhage ceased.

Case 6. Age 36. Operation Aug. 2, 1894. Interstitial fibroid. Profusely hemorrhagic and painful. Patient much reduced. Tumor three by five inches in diameter. Six months after operation much improved. No later report.

Case 7. - Age 35. Operation Nov. 11, 1893. Incipient interstitial bleeding fibroid. Two years after operation: Tumor reduced; hemorrhage ceased.

Case 8. Age 35. Operation Nov. 28, 1893. Painful, hemorrhagic interstitial fibroid, size of four months' pregnancy. Two years and four months after operation: Tumor much diminished; hemorrhage ceased: pain less but not entirely relieved.

Case 9. - Operation May 19, 1894. Interstitial, painful, hemorrhagic fibroid. Tumor size of four months' pregnancy. One year and ten months after operation: Tumor slightly diminished; hemorrhage materially reduced.

Case 10. -Age 35. Operation August 1894. Tumor interstitial four by six inches in diameter. Hemorrhage and pain excessive. Not much improved eight months after operation.

Case 11.--Operation November 7, 1894. Tumor interstitial, hemorrhagic, painful and size of four months pregnancy. One year and four months after operation: Tumor decreased in size and hemorrhage ceased.

Case 12. Operation Nov, 1894. Tumor interstitial, profusely hemorrhagic, painful, and three by six inches in diameter. One year and three months after operation: Tumor reduced ; hemorrhage less; pain not improved.

Case 13. Operation Feb. 1, 1895. Tumor intramural, two centers of development two inches in diameter each. Profusely hemorrhagic and excessively painful ; hemorrhage and pain ceased ; tumor nearly disappeared, as demonstrated by a laparotomy ten months later.

MINOR SURGERY FOR SUBMUCOUS FIBROIDS.

Pedunculated submucous fibroids may frequently be completely removed through the dilated cervix without interfering materially with the uterus. Unless the tendency to pedunculate is well established however, and the center of development comprising the tumor is the only center of fibroid development to be discovered in the walls of the uterus as shown by careful bimanual palpation, it should be treated by hysterectomy either vaginal or abdominal. An exception to this general rule would be when a pedunculated fibroid is discoverable either in the cavity of the uterus or hanging from the cervix with a long thin pedicle. In such a case the polypus should be carefully removed from the uterus, even though there were other centers of development to be discovered. The uterus as a whole, here, could be dealt with in a later operation if the removal of the pedunculated mass did not sufficiently relieve the symptoms.

The removal of an intrauterine pedunculated fibroid is usually a simple procedure. If the pedicle is small and long and the tumor is in a position where it can be easily reached with forceps, it may be grasped in a strong vulsellum and the tumor twisted until the pedicle is actually twisted in two. This can only be done with thin pedicles.· If the pedicle is broad the uterus should be sufficiently dilated (the patient under an anesthetic) to expose the pedicle, if it is necessary to accomplish this the cervix may be divided as high as the vaginal junction. The mucous membrane of the pedicle should next be cut in its entire circumference. Then the remaining portion of the pedicle composed of the blood vessels, connective and muscular tissue should be twisted in the same way that one proceeds to twist off a small pedicle. If the remaining portion of the pedicle is small it will give way by that treatment. If it is rather large and fleshy, after it has been twisted into a small bulk it may be grasped by a strong pair of curved pedicle forceps and the pedicle severed with scissors or a knife outside of the forceps. If the pedicle is very vascular the forceps may be left in place for six or twelve hours. If this does not seem necessary the forceps are removed and the uterus packed with iodoform gauze. If the forceps are left on the pedicle, gauze should be packed around them. The forceps may be removed in six or twelve hours without disturbing the gauze.

I do not favor attempting to enucleate a submucous fibroid of any considerable size if its principal bulk is buried in the walls of the uterus. Such a procedure is attended with considerable mechanical difficulty because of the position of the tumor in the cavity of the uterus; it is a difficult matter to secure hemostasis in such a location and finally one seldom reaches in such a procedure more than one of several centers of developments of the tumors which are situated in the uterus. In these cases a hysterectomy is more satisfactory.

A cervical fibroid developing toward the mucous membrane, if pedunculated, should be removed in the same manner as that described for removing a pedunculated intrauterine fibroid. A cervical fibroid of small size may be enucleated by incising its capsule, grasping the tumor with a vulsellum, and dissecting it from its bed. The cavity may be closed with buried antiseptic catgut sutures or it may be packed with iodoform gauze.

116

CURETTEMENT.

In many cases of hemorrhagic fibroids much of the hemorrhage and leucorrhea is caused by endometritis. A safe and oftentimes beneficial treatment for such cases is thorough dilatation of the uterine canal and curettement of its mucous membrane. While it will not ordinarily have a direct curative effect it will frequently relieve disagreeable symptoms for a long period of time. The dilatation should be gradual beginning with Goodell's dilators and afterward exploring the anterior of the uterus with the index finger to discover whether there are any projecting masses into the interior of the cavity. After thorough dilatation with the cervix exposed and grasped with small vulsellum forceps in order to steady the whole organ, a sharp curette should be made to traverse all portions of the endometrium. This should be accompanied with some form of antiseptic irrigation. The whole mucous membrane should be gone over at least three times with the curette, the canal then loosely packed with iodoform gauze, the vagina filled with the same and the patient put to bed for several days. The gauze should be removed in forty-eight hours. After that antiseptic vaginal douches must be given for several days

LECTURE VIII.

REMOVAL OF THE UTERINE APPENDAGES.

Battey, Tait and Hegar independently conceived, performed and contributed to modern surgery the operation of removal of the uterine appendages. In 1865 Battey "conceived" but did not publish "the idea of producing an artificial menopause for the remedy of disease." His idea was published in 1872. Hegar operated on the first case with the object of establishing an artificial menopause, a few days before Aug. 1, 1872, the memorable date of Tait's first operation for the same purpose. Battey did his first operation just sixteen days later, or Aug. 17, 1872. Thus the time was ripe and three great men of three great nations, separated by thousands of miles, discovered the fact, independently of each other, and shook the tree of progress which has resulted in such an abundant harvest.

The removal of the appendages for the cure of fibroids of the uterus is based on the facts: 1, that removal of the uterine appendages eradicates the part of the economy in which resides the organ or center of menstruation and produces an artficial menopause: 2, that removal of the uterine appendages accomplishes a reduction of the direct blood supply to the uterus and thereby produces atrophy by depleting the organ.

ARTIFICIAL MENOPAUSE.

It is yet an unsettled question where the exact seat

of control of menstruation is located. It is not the
province of this article to enter into the heated discussion as to whether this remarkable center lies in the
ovaries or whether it is situated in the nerve structure
of the Fallopian tubes. It is enough for us to know
that the menstrual center, wherever it lies, is eradicated, in the maximum proportion of cases, when every
vestige of both ovaries and Fallopian tubes is
removed. It is well that such is the case, because it
would be an awkward and incomplete operation which
would seek to leave either the ovaries or the tubes.
These two organs have a function which of necessity
is incomplete without both of them. It would be
difficult to remove the ovaries without interfering
with the circulation and position of the tube. It
would be equally impossible to remove the tubes
without interfering with the circulation and function
of the ovaries. Then, too, a much more secure and
desirable pedicle can be obtained when both organs
are included in a ligature than is possible if but one
of the organs is selected. Then, as the function for
which each of these small organs is designed is dependent upon both, and the removal of both is easier
and therefore safer than the removal of one of them,
and when we take into consideration the liability of
either of these organs to become diseased, if not
removed when opportunity permits, there seems to be
no further reason why both organs should not always
be removed when it is desirable to produce an artificial
menopause.

REDUCTION OF BLOOD SUPPLY TO THE UTERUS.

In a previous article we discussed the blood supply
to the uterus. We found that the organ depended
upon two sets of arteries for its nourishment, the uterine arteries and the ovarian arteries (Fig. 21). The
normal ovarian arteries are a trifle more than half the
size of the uterine arteries. They supply the ovaries,
the tubes, the fundus of the uterus and anastomose
with the uterine arteries which course along the sides

of the uterus giving off frequent horizontal branches
to the uterus. By referring to Fig. 21 it can readily be
seen that the ovaries and tubes may be removed without including the main channel of the ovarian artery.
Such a method of operating would deprive the
removal of the appendages for the cure of fibroids
of one of its chief features of benefit, viz., the reduction of blood supply to the uterus. For that reason
special care should be maintained by operators adopting this procedure to include in all cases the main
channel of the ovarian artery in their ligature. By

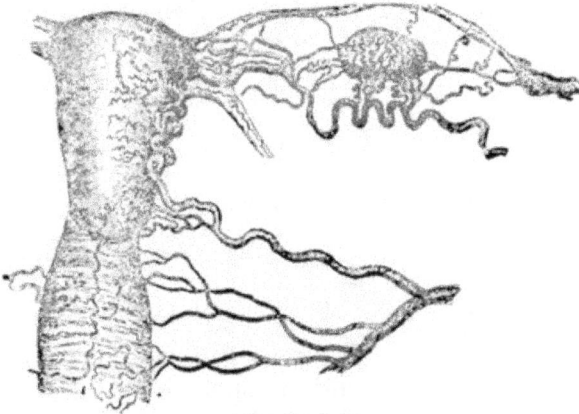

FIGURE 21.

tying this artery on both sides the large abnormally
developed uterus is instantly deprived of one-third of
its blood supply.

Dr. Byron Robinson, after witnessing my operation
of tying the broad ligament from the vagina, recognized the value of this principle of cutting off blood
supply to fibroid uteri, and afterward applied one of
the principles of my operation through an abdominal
incision, after having first removed the appendages, by
tying the uterine artery as it courses up the side of
the uterus to join the ovarian artery.

INDICATIONS FOR REMOVAL OF UTERINE APPENDAGES
FOR FIBROIDS.

But a few years ago this was the operation of selection for the relief of uterine fibroids when an operation of the severity of a laparotomy was deemed a necessity. It is seldom performed at present except as an operation of *dernier ressort*, when laparatomy has been instituted with the object of removing the tumor and uterus and, because of contraindications, the latter operation is found inadvisable. The reasons for this change of position are:

1. The operation of vaginal and abdominal hysterectomy has been so perfected that in patients of ordinary strength, with tumors without severe complications, the mortality of hysterectomy is not materially greater than that of double oöphorectomy.

2. The operation of removal of the appendages fails about three times in thirteen recoveries to materially reduce the size of the tumor, and fails in one case in thirteen recoveries to produce an artificial menopause; while hysterectomy on the other hand is absolutely sure of curing every case of fibroid of the uterus, which recovers from the operation both of hemorrhage and tumor.

This materially narrows the field of this operation which has done more to develop modern surgery than any other discovery of modern time, except the discoveries of Lister. The very enlightenment which it has created helps to make it obsolete. The operation now, in the hands of expert abdominal surgeons, is limited to cases: 1, where for some reason the operation is demanded because of prejudice against sacrificing of the uterus; 2, in cases where for some good reason quickness of time in operating is desirable; 3, in cases where unusual complications are revealed when the abdomen is opened which make hysterectomy impracticable; 4, in cases of small bleeding tumors in weak women who are near the menopause; 5, in cases of small hemorrhagic fibroids in weak women in whom laparotomy would not ordinarily be indicated but which are complicated with disease of the appendages.

THE OPERATION.

An abdominal operation is properly divided into five parts: Incision, removal of pathologic material, drainage, closure of incision and dressing.

The Incision.—After the skin has been prepared as described in Lecture VI, sterilized towels have been placed around the field of operation, the patient is thoroughly anesthetized with ether and the operating corps is in its place; the operator standing on the left side of the patient, with a sharp scalpel makes an incision from above downward in the median line, from about two inches below the navel to two inches above the upper margin of the pubis, an incision about three inches in length. This incision should be unhaggled and should extend in depth through the skin, superficial fasciæ, the fat between the superficial fasciæ down to the deep fasciæ which immediately covers the muscles. In experienced hands but one stroke of the knife is necessary for this. If the hemorrhage is but venous, sponges only are necessary to keep the wound dry. If there are any arterial points of bleeding they are caught in the points of forceps by the assistant. The operator by another stroke of the knife incises the white deep fasciæ, and if this incision is through the linea alba, the subperitoneal space is entered, as will be indicated by the bulging fat of this space. If the incision is to the right or left of the line the muscular coat of the abdominal wall will be exposed. The muscles are then separated by the handle of the scalpel in a stroke from above downward which brings into view the subperitoneal fat. The knife is carefully drawn over this from above downward, and this followed by a sweep of the scalpel handle separates the fat and subperitoneal tissue down to the peritoneum. The peritoneum is now caught in two catch forceps which are held up and separated laterally so as to present a sharp elevation of that membrane. This is carefully incised with the knife. When the peritoneum is opened it will be indicated by its sudden elevation in consequence of the entrance of air. An experienced operator will frequently open an abdomen carefully in thirty seconds. If it is done well it matters little if it takes five minutes. It is not necessary to seek for the linea alba if it does not happen to lie directly in the center of the superficial incision. It is more important that the wound should be direct and the different layers parallel than that the muscles should not be disturbed. When the peritoneum is opened between the forceps the index finger should act as a director above and below the opening and the peritoneum incised with scissors the full length of the wound, being careful not to wound the bladder below. Next attach the peritoneal edges of the wound at the center of the incision on either side to the deep fasciæ with small catch forceps. This prevents peeling off the peritoneum from the parietes in any subsequent manipulations.

Exploration.—With the index finger of the left hand I now make my exploration of the abdominal viscera including the appendages First the uterus is sought as a central landmark. From the uterus the finger is first swept to the left side along the Fallopian tube from the horn of the uterus. Just below the tube and above the broad ligament is the ovary. The opposite side is rapidly explored in the same manner. The exploration takes into consideration the size and position of the fibroid uterus, adhesions, the condition of the appendages, the possibility or feasibility of removing them and any abnormal developments.

Removal of the Appendages.—Our object, it must be remembered, in the removal of these organs must be to remove completely every vestige of ovary and tube and the thorough ligation of the main channel of the ovarian vessel. When the uterine tumor is not large, or if it has not developed into the broad ligament so as to spread out its folds and make it tense, it is an easy matter to ligate off the tube and ovary with one ligature. This is accomplished by lifting the tube and ovary with the loose broad ligament and making a pedicle of the infundibulo-pelvic ligament (Fig. 22 *a*), the ovarian ligament (*b*) and the Fallopian tube (*c*). The ligature No. 10 braided silk or No. 8 antiseptic catgut threaded in a round non-cutting needle is placed around the pedicle, never through it except as it penetrates and surrounds a portion of the infundibulo-pelvic (Fig. 22 *a*) and the ovarian (*b*) ligaments in order to prevent its slipping

FIGURE 22.

over the edge of the stump. If the ligature is allowed to transfix the pedicle at any other place than through the firm ligamentous tissues of the two ligaments mentioned, it is liable to produce venous oozing into the loose subperitoneal tissue beneath the constriction of the pedicle, resulting in small hematomas which frequently prove troublesome. After the ligature is placed the ovary and tube are drawn well up and the strand of silk or catgut is tied firmly, first with a double twist knot and then two single twists and the ends cut short (Fig. 23). A pair of snap forceps is then placed on the pedicle outside of the liga-

ture and the pedicle is severed about one-fourth
inch from the ligature, the stump cauterized or rendered sterile with some strong antiseptic, this antiseptic removed by a moist sponge and the pedicle
dropped.

If the uterus is considerably enlarged or if it especially develops into the broad ligaments, it is impossible to tie off the appendages of each side with a
single ligature. This is because the loose folds of the
broad ligaments of which the pedicle is ordinarily
constructed, have been occupied by the enlarged

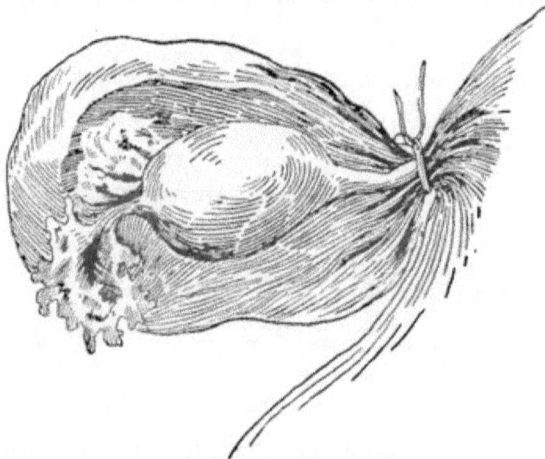

FIGURE 23.

fibroid uterus, and the ovary and tube are each flat-
tened out on the surface of the tumor and are held
fast by the peritoneum, which ordinarily acts as a
mesentery to each. Fig. 24. Under such a disposition
of affairs or any modification of it, the ovaries and
tubes should be tied off by first ligating the neck of
the tube as near the uterus as possible after anchoring
the ligature by a twist around the utero-ovarian ligament (Fig. 24 a); second by ligating the broad ligament outside the fimbriated extremity of the tube

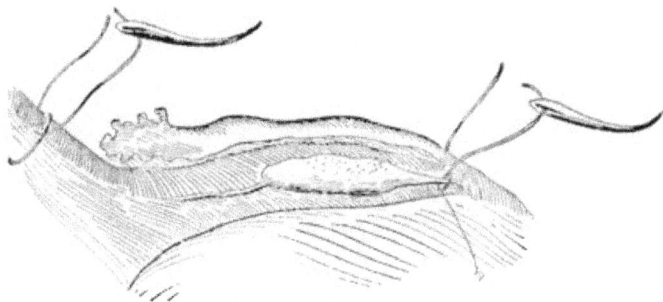

FIGURE 24.

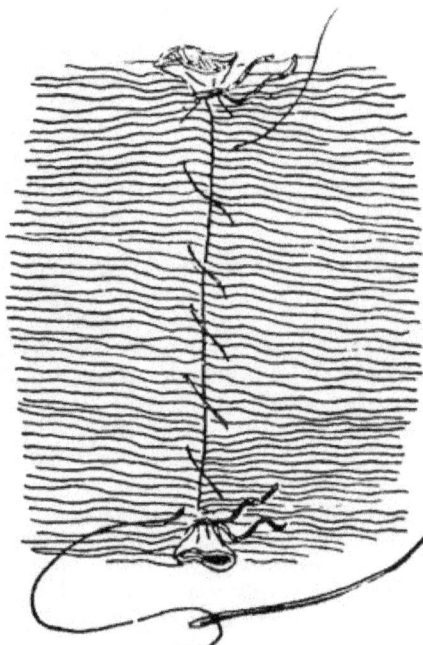

FIGURE 25.

deep enough to include the ovarian artery, anchoring the ligature by a twist around the infundibulo-pelvic ligament (Fig. 24); third, after removing the tube and ovary the two peritoneal edges representing the broad ligament to which the tube and ovary are attached between the two ligaments already placed, should be united from one pedicle to the other with a running stitch of fine antiseptic catgut. This makes a perfect exsection of the appendages, and leaves the peritoneum perfectly closed with no tension on either of the pedicles (Fig. 25).

DISEASED APPENDAGES.

All cases, unfortunately, are not typical like the ones we have described. We often meet with diseased appendages when opening the abdomen for the removal of these organs in cases of fibroids of the uterus. A pyosalpinx, or an ovarian cyst, or abscess of the ovaries are often encountered. Almost invariably when these things do exist, localized peritonitis has rendered them adherent to surrounding tissues, the uterus, omentum, intestines or the peritoneum of the broad ligament.

When these complications exist, and we have decided upon the removal of the appendages for the treatment of the fibroid, we have before us the problem of enucleation and excision of the diseased and adherent organs. The enucleation of an enlarged pyosalpinx, or an ovarian abscess, or a tubal ovarian abscess, or adherent appendages the result of an old peritonitis where pus is no longer present, is accomplished in practically the same way. When the abdomen is explored the abnormal condition of affairs immediately becomes apparent. Frequently there is an inextricable mass; occasionally the outlines of the tube and ovary can be traced and they simply appear as enlarged adherent semi-fluctuating cysts; while rarely, the appendages, not materially changed from their normal size, will be firmly imbedded and adherent.

I begin my enucleation in these cases, by passing

the index finger of my left hand, with the palmar
surface directed forward, down behind the isthmus of
the tube just as it is given off from the horn of the
uterus, hugging closely the body of the uterus until
I have reached Douglas's cul-de-sac. As a rule, at
this place I will find a line of cleavage as indicated
by the adhesions between the tube and the uterus and
the ovary, and the intestines and omentum posterior
to the ovary and tubes. This line of cleavage, which
as it gives way feels like two pieces of strong paper
which have been stuck together with fresh mucilage
giving way before pressure of the fingers between
them, can be followed rapidly, first with one finger,
then with two or more, until the whole adherent tube
and ovary are freed and lie ready to be ligated off in
the palm of the hand. This is the happy result in
the majority of cases.

In a few cases, especially after long standing disease,
the adhesions are strong and well organized. Here
great care and patience must be exercised, in order
not to go into an adherent bowel, or to the other
extreme and leave a portion of the stroma of the
ovary or a portion of the tube which may be just
enough to prevent the menopause and thus make our
operation a failure. Here, when the line of cleavage
fails to yield readily, it is well to place the patient in
the Trendelenburg position and separate the adhesions
after exposing them to sight. With this precaution
and the exercise of considerable care in manipulation
it is seldom that one need fail to accomplish an
enucleation.

When these adherent masses are once dug from
their beds the stumps, after ligation, should be rendered perfectly sterile by the application of the actual
cautery or strong chemic antiseptics. The parts
should then be dried, and where it is practicable the
raw surfaces should be covered with peritoneum.

Drainage.—If there is oozing from the raw surfaces caused by the enucleation, a glass drainage tube
should be placed in the cul-de-sac of Douglas, the

lowest point of the pelvis, while the toilet of the peritoneum is being completed and the abdominal sutures are being inserted. Before the abdomen is closed the tube should be pumped out in order to ascertain if there is more blood oozing from the peritoneal surfaces than would naturally be taken care of by the peritoneum. If more than a couple of drams of pure blood accumulates in the few minutes that are required to make the toilet and insert the sutures, the tube

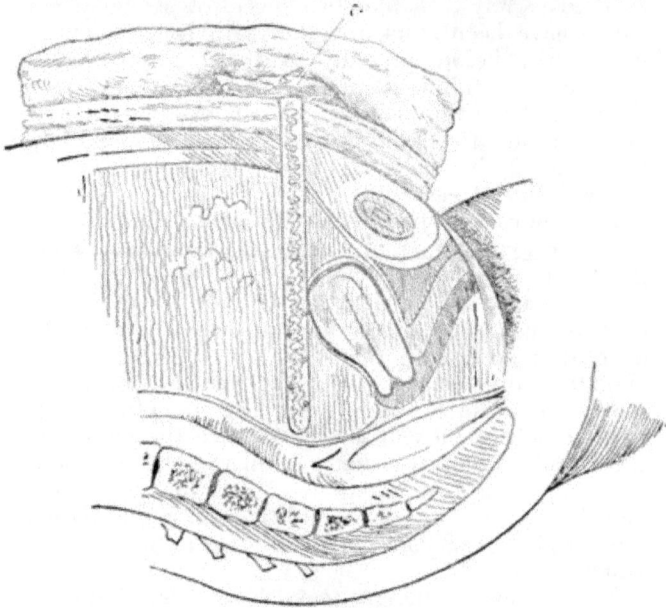

FIGURE 26.

should be allowed to remain, and the sutures tied so as to enclose it snugly. The tube should not be larger than an ordinary lead pencil, or about one-quarter of an inch in diameter. It should be long enough to project about one inch above the wound (Fig. 26) The abdominal dressings are then placed upon the wound around the tube and secured by

sterilized muslin bands which are pinned to broad
adhesive straps fastened to the sides of the abdomen
at some little distance from the wound. Over the top of
the tube is slipped a piece of rubber dam (Fig. 26 a)
about 12 inches square, the tube penetrating the center of the sheet rubber. The tube is then pumped
out with a small glass syringe with a long rubber
nozzle which will reach to its bottom (Fig. 27). After
finally emptying the tube a long narrow strip of sterilized gauze is carried to the bottom of it with a
straight metal sound, a small amount of the same
gauze is left as a loose dressing over the end of the
tube, over this is folded the rubber dam and fastened
with a sterilized safety pin, and over all this is placed
a liberal dressing of sterilized cotton and finally over
this a snugly applied bandage of sterilized cotton.

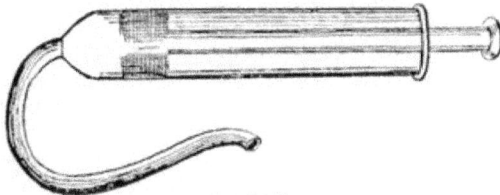

FIGURE 27.

Now when it becomes necessary to exhaust the drainage tube in an hour or two hours, it will not be necessary to disturb the wound dressing at all. The
bandage is unfastened, the layers of cotton parted in
the center and the rubber dam opened and spread out
on the cotton, the capillary drain of gauze removed
and the tube exhausted by means of the syringe, another strip of gauze inserted, the rubber dam refolded
and pinned and the external dressings readjusted.

I go into the detail of my method of caring for a
drainage tube in order to justify myself for using it.
As understood by the majority of good surgeons, it is
really a dangerous means of draining. But a few
weeks ago, in a discussion in a prominent society on
the subject of abdominal and pelvic surgery, I heard

a surgeon of no mean repute condemn the glass drain
and the suction pump, in unmeasured terms, when by
his very language, he demonstrated his ignorance of
the whole matter by saying that it was necessary to
uncover the abdominal wound every time that the
tube is exhausted. It is not necessary to uncover the
wound at all to dress a glass drainage tube. It is not
necessary to infect a glass drainage tube when uncov-
ering it to exhaust it with the suction syringe, if the
one doing the work is trained and competent. The
nurse should be well trained. She should wash her
hands to surgical cleanliness before loosening the
bandage or removing the cotton over the end of the
tube. She should rinse her hands in 1 to 1,000 bichlorid solution before unpinning the rubber dam; she
should again rinse them in bichlorid before removing
the capillary drain; she should take the glass syringe
with its rubber nozzle out of a pitcher of 1 to 1,000
bichlorid solution, rinse it quickly in sterilized water,
rapidly exhaust the tube, and eject the fluid into a
small glass graduate which has just been removed from
an antiseptic solution; when the tube is dry she should
take the steel sound out of a dish of bichlorid solution, a strip of gauze out of a fresh supply from a
sterilized package and insert it to the bottom of the
drainage tube. She should again rinse her hands and
then rapidly close the tube and replace the dressings.
Such a procedure requires two minutes if done by an
expert and intelligent nurse. I will agree with every-
body that this kind of drainage can not be carried
out by an ordinary nurse. But when well attended to
it is the most satisfactory method of keeping dry the
free abdominal cavity that we are as yet acquainted
with.

CAPILLARY DRAIN.

Occasionally after extensive enucleation of diseased
appendages we may be so situated that we have no
experienced nurse to leave in charge of a glass drainage tube; at the same time we *must* drain, and must
drain in a manner that the after-care of the drain can

be attended to by one of little experience. In these
cases combined capillary gauze and tubular drain,
through the Douglas cul-de-sac, may be resorted to.
A rubber tube one-quarter inch in diameter and about
twelve inches long, and a quantity of sterilized iodoform gauze in eight inch strips cut the strong way of
the gauze without knots, are selected. After insuring
thorough cleansing of the vagina the operator, guided
by an assistant's finger in the vagina, penetrates the
posterior cul-de-sac into the vault of the vagina with
a pair of sharp pointed scissors. The scissor blades
are then opened and between them from above downward, on a pair of forceps, is carried an end of the
strip of gauze and the rubber tube. The assistant
grasps these in the vagina and makes gentle traction
on them. The operator then loosens his grasp and
catches the abdominal end of the tube with a pair of
catch forceps. The tube is then drawn through until
from two to four inches of its upper end, according
to the extent of drain required, is left in the abdominal cavity. Enough of the gauze is drawn through
to make a loose packing for the vagina. A small bunch
of the gauze is left in the abdomen in the cul-de-sac,
around the end of the rubber tube. It is closely
packed so that it will remain in the position in which
it is first placed. The abdomen is then closed in the
usual manner. The vaginal end of the drain is arranged by cutting the tube off at the vaginal outlet;
and over the vulva and the end of the gauze is placed
a liberal pad of loose sterilized dry strip gauze. Orders
are given to change this outer gauze as often as it
becomes moist. When at the end of twelve to twenty-
four hours there is little drain, a portion of the gauze
may be withdrawn from the vagina, and if the drain
has been slight the tube may be removed at this time.
In twenty-four hours longer, if the drain is still small
or none at all, the drain may be completely removed.
After the gauze has been removed, a liberal vaginal
drain of gauze may be carried on the end of a forceps
to the vault of the vagina This may be removed in

twenty-four hours. After this nothing further is required but an occasional vaginal antiseptic douche. If drainage is profuse after twenty-four to forty-eight hours the gauze drain should be withdrawn more slowly.

Closing abdominal Wound.—I favor any method which will coapt all the tissues of the abdominal wound in the exact relation and to the same extent that they were originally. This can be accomplished by including all the tissues, skin, fat, superficial and deep fascia, muscle, subperitoneal fascia, and peritoneum in a row of silkworm gut sutures placed one-third of an inch apart. If I have some of my own specially prepared antiseptic catgut at hand, I frequently sew the peritoneal layer separately with a running thread of the gut and then include the remaining layers in the row of silkworm gut. This is especially desirable if one has a long wound in a hemorrhagic patient. It completely closes the abdominal cavity from any oozing from the abdominal incision. It also obviates the necessity of the silkworm gut sutures entering the peritoneal cavity, thus removing the remote danger of adhesions of the abdominal viscera to the points of peritoneum penetrated by the stitch, and danger of septic material gaining entrance to the peritoneal cavity along the route of the stitch, in case of external skin or mural suppuration. I am careful to include all the tissues of my wound in order that the abdominal walls after incision will be as thick at the wound line as at any other position. If it is not, there will be a concavity at this point on the peritoneal surface which will act as a point of resistance, and which will favor abdominal pressure on the wound and from which ventral hernia is more liable to result.

Before tying the silkworm gut sutures, I render the wound aseptic by washing thoroughly with 1 to 1,000 bichlorid solution (employing care that none of the poison enters the peritoneal cavity) and finally rinsing the wound with sterilized water. After tying the

main sutures of the wound I always put in superficial
stitches of fine silkworm gut wherever they are
necessary in order to insure coaptation of the skin
edges.

Dressings.—Sterilized iodoform powder mixed with
boric acid is dusted over the wound. Loose sterilized strip gauze is placed over the wound, and several inches around it, and over this is placed a dozen
thicknesses of sterilized sheet gauze. This is held in
place by sterilized muslin straps which are pinned to
broad bands of adhesive straps attached to the skin
on the outer borders of the abdomen. This prevents
the dressing from becoming displaced by any movements the patient may make, and it also supports the
wound and takes the strain off the sutures. Over this
is placed an abundance of sterilized cotton and over
the cotton in turn is placed a snug abdominal bandage
with a perineal T to keep it in place.

AFTER-TREATMENT.

For detail after-treatment I must refer the reader
to Lecture VI.

Dressing the Wound.—The wound is not disturbed
until the fourth day unless there are symptoms
indicating that it is not doing well, viz., pain, fever.
etc. At the end of the fourth day the nurse uncovers
the wound carefully, washes it thoroughly with
alcohol, and 1 to 2,000 bichlorid solution equal parts,
with sterilized cotton on the end of a dressing
forceps. It is then dried carefully and redusted with
sterilized iodoform and boracic acid. It is then recovered with fresh sterile gauze. On the seventh day
the same process is repeated and the stitches removed.
After that it is washed off in the same manner every
day until it is perfectly well.

ANALYSIS OF CASES.

I have removed the appendages for bleeding fibroids
of the uterus in 65 cases. These cases have all recovered from the operation. The history, subsequent to

the operation of a large per cent. of these cases, I have been unable to trace.

Cases 26, 28, 47, 48, 55, 61, 64, or 14 per cent. continued to menstruate indefinitely after the operation. Their symptoms were so severe in 26, 47, 48, 55, or 6 per cent. of the whole number that hysterectomy was afterward employed. In none of these cases was hysterectomy found necessary because of increase of the growth of the tumor. In the remaining cases, so far as I have been able to trace them, the tumors have reduced in size, hemorrhage has ceased and the patients have been materially benefited while in a small per cent. actual symptomatic cures were obtained.

LECTURE IX

HISTORICAL.

Vaginal hysterectomy is closely associated in its early history with cancer of the uterus. Greig Smith says, that it is probable that incision of the uterus was practiced by the ancient Greeks, but it is certain that it was subsequently forgotten. Soranus, in his book of " Diseases of Women," who lived in Rome a century before Christ, describes the operation as a surgical procedure for prolapsus. The first authenticated description of vaginal hysterectomy subsequent to this was given by Berengarius, of Bologna, in 1507

In 1560 Andreas A Crucé performed vaginal hysterectomy. J. Schenck a Grafenberg 1617 (Senn) relates a number of cases in which the uterus was removed through the vagina in whole or in part by ignorant persons who had not the faintest idea as to the difficulty or of the extent and gravity of the operation. In 1792 Saumonier removed an inverted uterus below a ligature. Hildanus 1646, Wrisberg 1785, Bernhard 1824 reported cases of accidental or unintentional removal of uteri by the vagina by careless midwives and others. Intentional removal of the uterus by surgeons have been reported by Zwinger, Vieussen, Baxter, Faivre, Alexander, Hunter, Joseph Clark and Jackson. (Senn.)

The real history of vaginal hysterectomy begins when it was deliberately planned and executed for the relief of definite pathologic conditions. To J. M.

136

Langenbeck in 1813 belongs the credit of opening
this page of history. He removed the uterus by
enucleation, using neither clamps nor ligatures and
his case recovered and lived many years. The postmortem demonstrated to his incredulous critics the
truth of his claim. Sauter, Jan. 28, 1822; Elias von
Siebold, April 19, 1823; Holscher, Feb. 5, 1824;
Elias von Siebold, again July 25, 1825; Langenbeck,
again Aug. 5, 1825; Recamier, July 26, 1829; Langenbeck, again Aug. 18, 1829; Roux, Sept. 20 and
Sept. 25, 1830; Recamier, again Jan. 13, 1830; Blun-
dell, Oct. 16, 1830; Siebold 1831, Delbach 1839 are
the bold pioneers who followed the lead of Langenbeck in Europe. From 1839 to the revival of
Czerny in 1878 there are no records of European
cases. In America, however, a few cases were put on
record during this long interval. Palmer Dudley
reports that Dr. John M. Esselman, of Nashville,
Tenn., in September, 1834, removed an inverted
uterus by means of the ligature, his patient recovering. This same surgeon repeated the operation successfully in August, 1843, for an inverted uterus
containing a fibroid. This is the first vaginal hysto-
rectomy for fibroids of the uterus I find recorded.
The first vaginal hysterectomy for cancer deliberately
undertaken and successfully executed in this country
was performed by Dr. Paul F. Eve, of Augusta, Ga.,
April 16, 1850. (Am. Journal of Medical Science,
1858.) Dr. L. C. Lane, of San Francisco, operated
for cancer Nov. 11, 1878, and at a later date in the
same year on a second case for cancers. Both cases
recovered. Lane executed this operation independently
of Czerny, who revived the operation in Europe by
performing his first operation April 12, 1878, or seven
months earlier than Lane.

From the revival of Czerny and Lane, with the
dawn of clean surgery, the operation of vaginal hysterectomy became a legitimate operation. In less than
twenty years it has made wonderful strides. It has
been performed thousands of times by hundreds of

137

operators, **and has,** undoubtedly, been the means of
adding many years to the sum total of human life.

VAGINAL HYSTERECTOMY FOR FIBROIDS.

It is argued that a patient suffers less real shock, on
an average, when submitted to a vaginal hysterectomy
than when operated upon by the abdominal route.
The only rational explanation that can be forwarded
to account for this fact is that the intestines and the
peritoneum are not subjected to exposure to the air in
the vaginal route, nor are they subjected to the handling which they are liable to receive in the abdominal
operations. While in our improved methods, the
abdominal contents are exposed and handled to a
small degree compared to former times, at the same
time we can not help but recognize that there is less
shock after a perfect vaginal hysterectomy than after
an abdominal hysterectomy in cases of like severity.
A vaginal hysterectomy avoids the abdominal scar,
which so many patients dread as a brand of mutilation which must be carried through life after all
abdominal operations. Many patients I find have
this wholesome dread to such a degree that there
seems to be no comparison in their minds between
the two operations. An abdominal operation contains all the horrors of a most dreaded affair, while a
vaginal operation with no sign of mutilation left, is
contemplated like a normal labor with dread but
resignation. An abdominal scar, it is true, will frequently become the seat of considerable irritation and
rarely the seat of severe neuralgic pains. There is
always, too, the remote possibility of ventral hernias
developing in an abdominal scar. It is also claimed,
by not a few operators, that safer and more satisfactory drainage can be obtained, when it is required,
through the vaginal route than by the abdominal.
However, as soon as we undertake to do a vaginal
hysterectomy on anything but the smallest kind of a
fibroid, we are hampered by the narrow limits in
which we have to do our work, and, therefore, if the

tumor is of considerable size, the extra time its proper
removal from the vagina requires, off-sets what is
gained by non-exposure of the abdominal viscera.
So that the rational surgeon must discriminate here,
as everywhere else in surgery, and select the operation which best suits the individual case. If he has
a small fibroid, or a large fibroid with relaxed ligaments and a large, roomy vagina, he should select the
lower route; whereas, if he has a large fibroid high in
the pelvis, or a small one with a narrow, contracted
vagina and rigid tissues he should do a laparotomy
and remove the tumor from above.

Methods.—Vaginal hysterectomy for fibroids may
be divided into two grand divisions: 1, removal of
the uterus and its fibroid masses as a whole without
division, or vaginal hysterectomy proper; 2, removal
of the uterus and its accompanying fibroid development in piecemeal or morcellement.

1. VAGINAL HYSTERECTOMY PROPER.

Indications.— Vaginal hysterectomy proper for
fibroids must of necessity include only the smallest
tumors, or at best fibroid uteri with long, slender subperitoneal projections. The operation is often the
ideal method of treating small multiple fibroids, which
are so frequently the seat of severe uterine pain and
excessive hemorrhage. Fibroids of considerable size
may frequently be treated by vaginal hysterectomy,
when the uterus is low in the pelvis and the vagina is
large and the tissues loose. It is an easy matter to
turn a complete vaginal hysterectomy for fibroids into
a morcellement should any unlooked for enlargement
manifest itself.

TECHNIQUE OF VAGINAL HYSTERECTOMY PROPER.

The patient should be prepared with the same care
and manner that I have described in Lecture VI on
preparatory treatment for laparotomy. Special care
should be maintained to render the vaginal tract and
external genitalia aseptic. The patient should be

anesthetized with ether. She should be placed in the
exaggerated lithotomy position with the limbs supported by some mechanical device which will hold
them firm and for any required length of time. Otherwise, they should be supported by strong skilled
assistants on either side, who will also hold the vaginal
retractors. Immediately before the operation begins
a nurse or a third assistant should thoroughly scrub
the external genitalia and the vagina with green soap.
This should be thoroughly rinsed off, and the parts
should be thoroughly washed with 95 per cent. alcohol and finally rubbed with 1 to 1,000 bichlorid of
mercury and then douched off with sterilized water.
Moist sterilized towels are placed around the field of
operation. The operator takes a seat at the foot of
the table on a small stool. At his right are his instruments with the surgical nurse to do his bidding, At
his left is the nurse who superintends the irrigator of
sterilized water and handles the sponges.

Operation.—Two small vaginal retractors, with short
broad blades, are introduced and held by the assistants so as to retract the anterior and the posterior
vaginal walls and expose the cervix uteri. The cervix
is grasped by a light pair of vulsellum forceps with
four teeth, the uterus rapidly dilated by first introducing a small dilator and then a large strong one,
until its interior can be reached and thoroughly
explored with a sharp curette. The uterine cavity is
thoroughly curetted and then rendered aseptic by washing out with a solution of 1 to 100 bichlorid of mercury.
It is then loosely filled with sterilized gauze. Through
the cervix. by means of a curved needle, is passed a
strong double handling string of braided silk, and
this is tied over the os uteri in such a way as to close
the canal. The vulsellum forceps are now removed
and the strong silk ligature is henceforth employed as
a means of handling the uterus. The uterus is now
drawn well down by making strong tractions and the
cervix drawn back so as to expose the anterior utero-
vaginal fold. With a curved scissors the mucous

membrane of the vagina at its attachment to the
uterus anteriorly is penetrated and the incisions carried to the right and to the left following the utero-
vaginal junction, until the incisions meet posteriorly
and the uterus is completely severed from the vault
of the vagina (Fig. 27). The assistant now grasps the

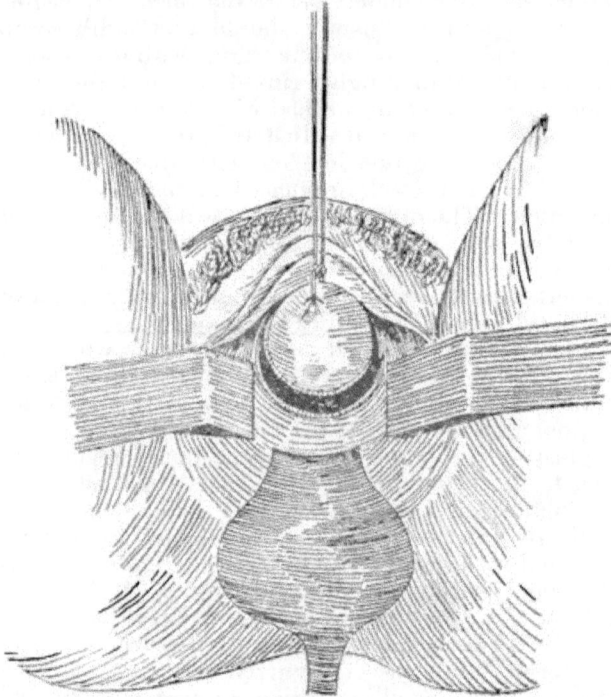

FIGURE 27.

handling string and makes downward and backward
traction, while the operator with the index fingers of
both hands carefully separates the bladder from the
anterior surface of the uterus. If there are any firm
bands connecting the two organs, they should be severed with scissors near their uterine attachment, always

keeping the point of the scissors against the firm
uterine tissue. As soon as the utero-vesical fold of
the peritoneum is reached with the fingers the two
fingers should be separated laterally, so as to detach
the bladder from the anterior surface of the broad
ligament, and also for the purpose of pushing the
ureters, which pass under the broad ligaments near
the cervix, well to the sides of the pelvis.

The assistants now draw the cervix forward and the
operator separates the uterus from its posterior
attachments and the two fingers penetrate through
the peritoneum into Douglas' cul-de-sac. The fingers
are then separated laterally tearing the peritoneum in
that direction. A large dry gauze sponge, with a
string attached, is pushed through this opening and
spread out above the uterus. The broad ligaments
and the appendages are then rapidly examined. The
peritoneum in front of the uterus between it and the
bladder is now torn through and the broad ligaments
are the only attachments left between the uterus and
the patient. If the uterus is not too large and the
broad ligaments are loose and the vagina large, one
pair of strong forceps will secure each broad ligament.
The uterus is drawn well down and the operator slips
his index finger of the left hand behind the left broad
ligament and crowds the appendages toward the
uterus until he can hook the finger over the ligament
outside of the appendages. With the uterus held
well down and steadied by one of the assistants, the
other assistant holding the bladder well out of reach
by a long narrow bladed retractor, the operator with
his right hand slides a strong pair of Byford's clamp
forceps (Fig. 28) over the broad ligament, the posterior blade following the lead of the index finger,
which is still holding the ligament, until they include
its whole width, and project half an inch beyond its
upper edge when they are closed and locked. The
jaws of the forceps should be examined carefully to
see that they include all the tissues necessary, and
that it compresses all portions sufficiently. The locks

of the forceps should be securely tied. With the
index finger guiding the scissors, the clamped ligament is now severed close to the uterus. If the
uterus is not too large and the right broad ligament
is long the organ can be delivered as the next step
and when delivered the right broad ligament may be
clamped with ease outside of the vulva. If this is
possible the clamp should be applied outside of the
appendages and the uterus cut away. Frequently,
however, the uterus can not be delivered until the
other clamp is applied and the ligaments severed.
Under such circumstances the forceps should be carefully applied exactly like the first one and the ligaments divided with the scissors from below upward

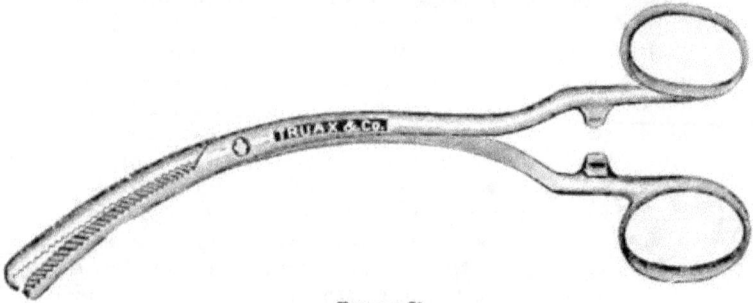

FIGURE 28.

while the assistant makes slight traction on the uterus
until the organ is free, when it is delivered. The
broad ligament forceps are carefully examined now,
to be sure that each is doing all the work required of
it, viz., including the whole ligament in its grasp and
firmly compressing every portion sufficiently tight to
maintain hemostasis. Should any portion need rein-
forcing, a small pair of straight hemostatic forceps
may be applied to the projecting free end of the severed tissue. Occasionally it is not practicable nor
safe to include the whole broad ligament in one pair
of forceps because of its width and bulk, while again
it may be difficult to place the forceps on the whole

ligament at once, because of a too narrow vagina
or a highly situated uterus with short ligaments.
Here the bulky base of each broad ligament
should be clamped first, with short stout catch
forceps and the ligaments severed up to within
a short distance of the forceps' bite. The uterus then
can be drawn down and the remaining portion of the
broad ligaments can be secured in one pair of forceps
on each side. The last forceps are placed on the
uterine side of the first pair. All the forceps are now
held by the assistants to their respective sides of the
vagina with their handles separated in such a manner
as to act as lateral retractors. The sponge is removed
from the pelvis and the toilet of the peritoneal cavity
is made by drying it with sponges on holders. The
posterior retractor is now inserted and the operator
seeks the edge of the peritoneum which covers the
bladder, grasps it with a catch forceps and draws it
down and with a running stitch of antiseptic catgut
attaches it to the upper end of the anterior vaginal
wall. An anterior retractor is now inserted and the
edge of the peritoneum covering the rectum is attached
to the upper end of the posterior vaginal wall in the
same manner. This insures hemostasis of the anterior
and posterior vaginal edges, and covers an otherwise
uncovered gap of connective tissue space.

Drainage. -The forceps are widely separated, two
narrow retractors hold open the vagina anteriorly and
posteriorly, a square piece of sterilized iodoform gauze
two feet wide is placed with its center over the vulva.
and with a large pair of dressing forceps its folded
center is carried well into the vagina beyond the ends
of the forceps, so as to form a bag. It is then loosely
packed with strips of sterilized iodoform gauze and
the edges of the filled bag are left projecting several
inches from the vulva. It is folded over the vulva.
The handles of the clamp forceps are wrapped in
gauze. A liberal supply of absorbent cotton is placed
over and around the forceps and over the perineum
and vulva, and all held in place by three small perin-

cal bandages, one passing between the handles of the forceps and the other two outside of the forceps handles.

VAGINAL HYSTERECTOMY BY MORCELLEMENT.

Indications.—Vaginal hysterectomy by morcellement may be done for fibroids of considerable size, the limit of maximum size on which the operation may be safely undertaken depending on the skill and experience of the particular operator. The writer does not favor the operation where the uterus is too large to deliver easily after bisecting, preferring to undertake such cases by the abdominal route. The operation is now performed every day, however, by an increasing number of skillful men on fibroids of every size, even on tumors reaching well above the umbilicus.

Polk, a firm believer in morcellement for fibroids, lays down the following indications: 1. Whenever the mass is largely within the pelvis, especially if it is fixed therein by adhesions. 2. Whenever the mass is soft and, therefore, compressable as in myoma and fibrocystoma. 3. In all other cases where we have a patient in good condition whose pelvis is shallow, where the vaginal canal is roomy, and in whom the evidence of a pyosalpinx above the pelvis brim are absent. Péan, Segmond, Richelot, Jacobs, Henrotin and others do not make suppurating appendages a contraindication to this method of operating.

TECHNIQUE.

Cases with Uterus only Double its Natural Size.—In these cases the technique is very similar to that for simple vaginal hysterectomy, with the exception that the uterus is bisected with an antero-posterior incision.

Step 1: The vagina is severed close to the cervix, as in ordinary vaginal hysterectomy, and the uterus denuded until the posterior and anterior cul-de-sacs are opened.

Step 2: Grasp the anterior lip of the cervix on either side with strong bullet forceps or two well embedded

handling strings, and making strong traction split the
anterior wall of the uterus with strong scissors with
the posterior blade guided by the uterine canal
(Fig. 29). When the scissors have reached the limit
of exposure of the uterus the edge of the split uterus
at the highest point of the incision should be grasped
by the bullet forceps, and with this new grasp the
uterus should be drawn down still farther and the

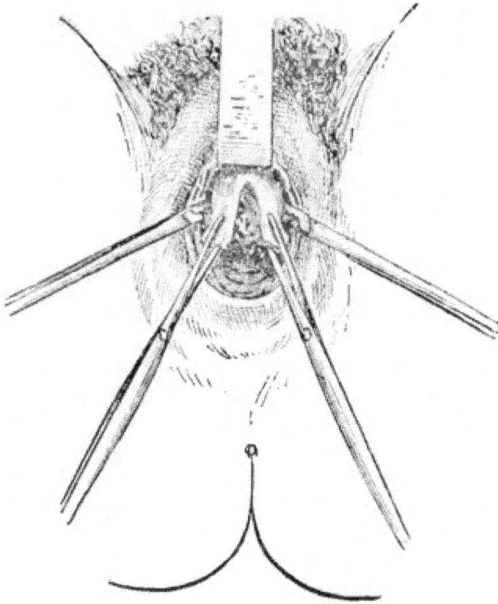

FIGURE 29.

splitting process continued. When the uterus is
usually movable or the broad ligaments unusually
long, sometimes at this point the partially split uterus
is completely anteverted and the fundus is delivered.
As a rule, however, whenever it is necessary to split
the uterus at all in order to remove it, the bisecting
must be carried to completion.

Step 3: When the bisected uterus is well drawn down, the increased movability of the organ in consequence of being in two pieces enables one to clamp the broad ligaments with one, or at most two forceps, and the respective halves are removed. After one side has been cut away it is an easy matter to clamp and remove the opposite half.

Step 4: Finish operation and apply drainage as in simple vaginal hysterectomy.

Variations of Procedure.—If it is convenient it is often better to attempt to clamp the broad ligaments immediately after opening the two cul-de-sacs, in order to save the patient as much blood as possible. When it is not possible to clamp the whole broad ligament, the base of the ligaments with the uterine arteries may be secured. As a preliminary every precaution should be observed to render the cavity of the uterus aseptic.

Cases with Uterus More than Double Its Normal Size.—In these cases the uterus must be removed by piecemeal. In order to accomplish this so as not to make a horrible failure, a thoroughly systematic course must be observed by one skilled in the details of pelvic surgery and surgical emergencies. No two cases are alike. Consequently no two operations are ever identical.

Step 1: Circular incisions around the cervix, after first grasping the anterior and the posterior lip of the cervix with strong forceps. The uterus is denuded anteriorly and posteriorly and the posterior cul-de-sac is opened. An attempt is then made to enter the anterior cul-de-sac.

Step 2: Clamp forceps are now placed on the base of each broad ligament high enough to include the uterine artery and its branches, and the ligaments are cut nearly as high as the point of the clamp.

Step 3: The cervix is split into halves by a lateral incision on the line of the uterine canal with strong scissors or a knife (Fig. 30).

Step 4: With the anterior lip well drawn down and

firmly held, the posterior lip is drawn well down and amputated. The remaining stump of the posterior half of the uterus is firmly grasped in forceps, keeping the uterus well down in the field of operation.

Step 5: If the uterus is not too large at this point a single clamp forceps may be placed on the remaining portion of the broad ligament.

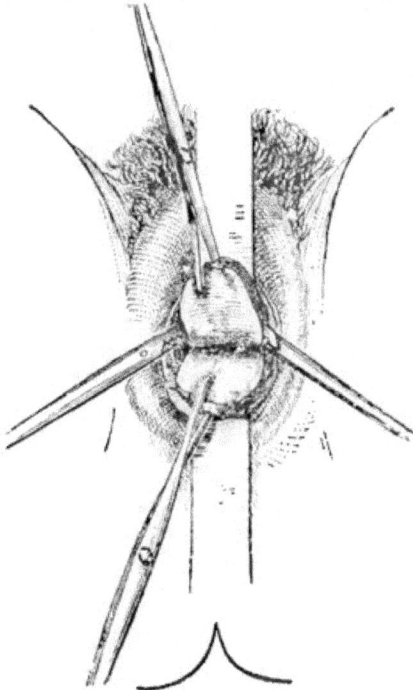

FIGURE 30.

Step 6: With hemostasis well secured, from this point on the splitting of the uterus is continued and morcellation is proceeded with by amputation of first one-half or a portion of a half and then the other, severing the broad ligaments by degrees until by piecemeal the whole uterus is removed. Care should be maintained to have a secure hold of the uterus with forceps at some point besides the point of amputation at all times, in order that it may not slip out of the field of operation. As soon as the uterine mass has been reduced sufficiently so that it may be delivered it should be removed. Care must be exerted to secure all large subperitoneal masses which from their situation might by carelessness be accidently separated from the tumor and escape beyond the reach of the finger or forceps. As the morcellement progresses a finger in the posterior cul-de-sac from time to time learns facts of value to the operator.

Remarks.—The care of the forceps and the care of the vaginal and peritoneal edges, after this modified operation, is identical to that after the simple operation. The toilet of the operation field and the drainage is the same. If adhesions exist they must be carefully separated. If the anterior cul-de-sac is elevated so that it can not be entered before amputation is commenced, amputation should be proceeded with and the uterus gradually drawn down until the cul-de-sac can be opened.

Accidents to be avoided.—Wounding of the bladder or intestines and clamping one or both carelessly are accidents which must be carefully guarded against in vaginal hysterectomy of any kind. Severing of important blood vessels before they are securely clamped is another annoying accident, because of the tendency of the unsecured blood vessels to retract into the loose connective tissue of the broad ligament, where they will continue to bleed out of reach of hemostatic forceps. To avoid this accident great care should be observed to securely clamp all tissues before severing, and if there is the slightest doubt about the security of any portion of the divided ligament after it has been cut, a second clamp should reinforce the first. To avoid clamping the ureters the forceps should not be applied to the base of the broad ligament until the bladder is thoroughly separated from its anterior surface, until the finger can sweep between it and the broad ligament to the sides of the pelvis. This insures the pushing of the ureters out of the reach of the forceps. This same maneuver insures the integrity of the bladder also. If the bladder is found so adherent at any portion that it is not readily separated with the finger, scissors should be employed to dissect it from the face of the uterus or tumor, great care being observed to avoid wounding the bladder with the scissors by keeping the point of that instrument against the uterus. By following the imperative rule of separating the bladder early and keeping it out of the field of operation by the use of an anterior retractor, it will never be wounded. To avoid wounding the rectum the same care should be observed in entering the posterior cul-de-sac as is exerted in opening the anterior one. If one carelessly opens this pouch, it is an easy matter to strip the peritoneal covering of the bowel posterior to the cul-de-sac and miss entering the peritoneal cavity entirely; while by ignorantly pursuing this false track the rectum may be penetrated. To avoid this embarrassing predicament *stick to the uterus.* If one does strip off a portion of the uterine peritoneum it will do no harm and the rectum is safe.

USE OF LIGATURES FOR VAGINAL HYSTERECTOMY.

By some operators ligatures are employed for securing hemostasis instead of clamps. In simple, uncomplicated cases ligatures may be employed with ease. In morcellement, where high and excessive manipulation is required the ligatures are impracticable because of the difficulty of applying them, and because of the difficulty of preventing them from becoming loosened by the subsequent manipulation of the parts. When ligatures are employed they prolong the convalescence if they are left long and allowed to ulcerate away. The time required for the accomplishment of that act is from twelve to forty days. During all this time it requires great diligence on the part of the attendant to prevent infecting of the ligatures; in fact, it is seldom prevented. An offensive vaginal discharge bears evidence of the fact, until the ligatures are finally discharged. If ligatures are employed and cut short, with the idea of burying them, it frequently happens that they become infected, even when the greatest care is observed to preserve cleanliness. The reason for this is the necessity for drainage in almost all of these cases. The method of drainage makes the wound practically an open one. Hence the danger of some portion of the otherwise buried ligatures becoming infected. Once infected long months of pus discharge from vaginal fistula is the sequel. This is all avoided when hemostasis is secured by strong forceps, because the means of hemostasis (the forceps) are removed in forty-eight hours, and nothing is left of a foreign nature which may become infected.

AFTER-TREATMENT OF VAGINAL HYSTERECTOMY WITH CLAMPS.

Shock is treated on the lines laid down in Lecture VI. I must, also, refer the reader to that Lecture for the detail treatment of the bowels and care of the patient as regards drink, diet. etc. The bladder is emptied every eight hours with a catheter until the forceps and the first drainage is removed. The catheter should be employed oftener if it is necessary. The nurse employs an aseptic glass catheter with a small nozzle which will run the urine off into a bottle. The urethra is carefully exposed before the catheter is introduced and thoroughly wiped off with a saturated solution of boracic acid.

Forceps.—The locks of the forceps are carefully tied at the time they are put on to avoid accidental unclasping of their blades. The handles are kept covered with sterilized gauze. If·it gets soiled at any time it is changed. At the end of twenty-four hours all but the principal forceps are removed. The string securing the lock is cut, the lock unfastened carefully, the blades opened enough to loosen their grasp on the tissues and then they are carefully removed. At the end of forty-eight hours the main forceps are removed in the same manner. If the tissues in any particular case showed unusual tendencies to bleed at the time of the operation I allow the forceps to remain twenty-four hours longer. When the large forceps are removed they should be opened widely before an attempt is made to withdraw them, and then they should be brought out with a backward motion in order to avoid catching the tissues with the projection on the posterior blade.

Dressing.—If the external portion of the drain becomes soiled with accidental discharges of urine or excessive drain fluid it should be removed and replaced with a new external pad of loose sterilized gauze as often as is necessary. Twelve hours after the last forceps are removed about one-third of the external gauze drain should be removed and a fresh pad of gauze placed over the vulva. In twenty-four hours another third should be extracted, and in twenty-four hours more, or sixty hours after the last forceps are removed, all of the balance should be taken out. When the last gauze is removed an external irrigation should be employed of 1 to 5,000 bichlorid solution followed by a plain water irrigation. A small sterilized iodoform drain should now be carried carefully about three inches into the vagina on a strong pair of dressing forceps, and the end of the drain allowed to protrude from the vagina. Over this is placed an antiseptic absorbent pad. In twenty-four hours this drain is removed and then eighty-four hours after the last forceps are removed, and the peritoneum has had ample time to close, the first vaginal douche is given.

Douches.—This douche must be given with extreme care by a nurse who understands all the responsibility she is entrusted with. The douche point must be made of glass, bulbous, with openings directed only at right angles to it. The patient should be placed on her back at the edge of a bed with feet supported on two chairs. A Kelly pad should be under her buttocks. The reservoir containing sterilized water of a temperature of 105 should be placed but eighteen inches above the patient's hips, in order to have but slight pressure. The nurse after thoroughly preparing her hands, inserts two fingers into the vagina about two and one-half inches, and between the fingers extending to within one-half inch of their extremities is inserted the douche point. The water is turned on with every precaution employed to secure immediate and free return current. The douche is repeated in this way, the nurse introducing the fingers and douche point a little further each time until the vault of the vagina is reached, every twelve hours until the sixth day from the removal of the forceps, when 1 to 5,000 bichlorid of mercury solution may be substituted for the plain douche, always following the bichlorid douche by a plain one. It must be obvious why I insist on the great care in employing this douche. The peritoneal cavity is expected to close in a few hours after the gauze is removed. Frequently, there is no doubt, it is closed off a few hours after the operation is finished. Notwithstanding this tendency to early closure of the peritoneal cavity, carelessness in employing the first few douches, if free return stream is not provided and great pressure employed by placing the reservoir too high, might result in breaking up the union of the tissues and fill the abdominal cavity with the fluids and *débris* of the vaginal tract. After each douche the vulva should be covered with an antiseptic pad. After each urination or movement of the bowels the external parts should be douched off with sterilized water and the antiseptic dressing renewed.

Getting Up.—Patients manifest a desire to get up earlier after vaginal hysterectomy than after abdominal operations. There is less prolonged reaction in the way of nervous exhaustion as there is less immediate shock with the vaginal operation. I allow my patients to begin to sit up on the twelfth to fifteenth day. They can leave the hospital from the twentieth day onward.

Results.—I have performed vaginal hysterectomy for uncomplicated fibroids of the uterus forty-one times: one death resulted from the operation. At least twenty of these cases were performed with clamps as the exclusive means of securing the broad ligaments. In the balance of cases ligatures of silk alone or ligatures reinforced with forceps were employed. The majority of my cases were small tumors. In two cases there was a troublesome hemorrhage within twelve hours after the operation, from the vaginal edges, which was, however, easily controlled by applying small catch forceps to the bleeding point. I opened the bladder in one of my early cases; it was subsequently closed in a secondary operation. I have never had secondary hemorrhage or the slightest oozing after removing the clamps. I have never had a vaginal hernia. I have never had a troublesome vaginal fistula. In one case where ligatures were used and left long they did not come away for nearly six months.

Dr. Edward Garceau of Boston, to whose writings on vaginal hysterectomy I am indebted in preparing this article, gave in his excellent paper on "Vaginal Hysterectomy as Done in France," in the *American Journal of Obstetrics*, Nov. 3, 1895, the following table of operations with their results, to which list I add my own cases. A list like the subjoined can not be said to represent the average mortality of the operation because it includes the work alone of the most experienced operators in this line in several countries. At the same time it must be remembered that this represents pioneer work in a comparatively new procedure.

Operator.	Cases.	Deaths.
Mayer	1	0
Péan	200	4
Jacobs	22	2
Mangiagalli	8	0
De Ott	100	0
Carle	22	0
Calderini	1	0

Bockel	3	0
Routier.	6	0
Richelot.	43	1
F. H. Martin	41	1
	417	8

This gives a mortality of 1.7 per cent.

LECTURE X.

HISTORY.

Dr. Gilman Kimball of Lowell, Mass., was the first to deliberately plan and execute an abdominal hysterectomy for fibroids of the uterus. The operation was performed in August, 1853. On June 25 of that year, Dr. Walter Burnham of the same city removed a portion of the uterus for this disease. Ten years later, December 19, 1863, Koeberle did his first hysterectomy for fibroids, with external fixation of the stump of the uterus. He employed a metallic ligature with a special device for tightening it. Péan soon followed Koeberle and supplementing the latter's work by the free employment of forci-pressure, and by publishing a systematic technique, which included the employment of steel pedicle pins over the metallic ligature for maintaining the pedicle extra-abdominally, his name became inseparably associated with hysterectomy by the extraperitoneal method. Dr. M. M. Latta, of Goshen, Ind., completed an abdominal hysterectomy by tying the broad ligaments in sections down to vagina July 6,1876. The elastic ligature for temporary ligation was first employed by Kleeberg, of Odessa, July, 8, 1878. In August, 1878, Martin recommended the provisional elastic ligature. Hegar about the same time recommended the elastic ligature for permanent intraabdominal ligation of the pedicle.

In the evolution of abdominal hysterectomy many methods have been adopted and at different times each have had their advocates. The yearning for perfection has made confusion, owing to great efforts in many directions. In the early history of the operation the best results came with the extra-peritoneal method of treating the stump. This continued until the last few years when, with improved technique, greater experience, and to avoid unpleasant sequela the pendulum is irresistibly swinging toward the intra-abdominal pedicle. The history carries us through the following methods: 1, extra-peritoneal or Péan's; 2. intra-peritoneal or Schroeder's; 3. complete hysterectomy or Eastman's; 4, vaginal fixation or Byford's; 5. ligation of arteries outside of uterine tissue with intra-peritoneal stump or Stimpson-Baer method.

1. Extra-peritoneal Method.—The extra-peritoneal method as originally carried out by Péan, consisted in clamping the neck of the tumor with a wire clamp or serre-nœud including the broad ligaments with the appendages, preventing slipping of the constrictors with pedicle pins, excising of the tumor, fixation of the pedicle in the lower angle of the abdominal wound, and closure of the abdominal wound closely down to the stump. Joseph Price of this country has improved this method until it is well-nigh perfect. His success with it has been phenomenal.

Hegar and Kallenbach modified it by carefully securing the pedicle after Schroeder's method and uniting it in the abdominal incision extra-peritoneally but beneath the closed incision.

Kelly and independently Van de Walker modified it by making temporary fixation, like Péan, until liability to hemorrhage had ceased, when the wire was removed and the pedicle allowed to contract into abdominal incision.

2. Intra-peritoneal Method.—This, as practiced by Schroeder, consisted primarily in constricting the pedicle with the Kleeberg rubber band, removing the tumor, paring down the stump, taking from its center a wedge-shaped piece of the bulky tissue, cauterizing the canal, closing the stump by strongly sewing together the edges of the wedge-shaped incision and finally sewing over all the peritoneal edges. The stitching of the stump was intended to be secure enough so that all subsequent oozing was made impossible after the final removal of the rubber ligature. The pedicle was then dropped, as is the pedicle after ordinary ovariotomy, and the abdomen closed.

This method was modified by Olshausen, Charles T. Parkes, Zweifel, Hofmeier and others.

(a) Olshausen modified by securing the pedicle with a rubber ligature, and sinking the whole by sewing over it the peritoneum.

(b) Charles T. Parkes modified it by ligating firmly with strong silk and cauterizing the tissues of the pedicle to firm bone-like condition with the actual cautery over a temporary clamp.

(c) Zweifel tied the pedicle firmly with a strong multiple ligature of silk, securing it in this manner in several parts.

(d) Marcy of Boston, 1881, secured an intra-abdominal stump by sewing from the outer edge of one broad ligament to the other with thirteen cobbler's stitches; including in the process ovarian arteries, broad ligaments, uterine arteries and the stump of the uterus formed by the cervix uteri.

(e) Hofmeier carefully ligated the pedicle in its circumference without closing the cervical canal, and closed its abdominal end by covering with peritoneum. Drainage could take place into the vagina through the patulous canal.

(f) Goffe and Albert independently employed treatment similar to Hofmeier's, with the addition of applying a capillary drain through the open cervix into the vagina.

3. *Complete Removal, Eastman's methods.*

(a) In 1888 Dr. Mary A. D. Jones removed the entire uterus, including the cervix, by employing long hemostatic forceps for the lower portion of the broad ligament, and severing the cervix from the vagina.

(b) Joseph Eastman's method, 1889: The broad

ligaments are tied off, including the appendages, the vagina opened posteriorly by elevating it by means of a special staff constructed for the purpose, which is held by an assistant, the vaginal edges are ligated with long ligatures which afterward serve to invert the edges into the vagina, and the cervix and stump are progressively cut away. The peritoneum is sewed over the inverted vaginal edges, the abdominal wound is closed, and the vagina packed with gauze. The mass of the tumor, if cumbersome, may be cut away, previous to opening the vagina, by putting on a temporary rubber ligature.

(c) Eastman in 1884 enucleated the stump without first tying the uterine arteries by peeling the pedicle portion of the uterus with a serrated gouge, keeping inside of the uterine arteries in their course up the side of the uterus.

4. *Vaginal Fixation, Byford's Method.*—The broad ligaments are tied with silk and severed. The cervix is secured with provisional rubber ligature, the tumor cut away, the pedicle firmly tied with multiple silk ligatures, left long, the stump trimmed and closed with long silk ligatures, an opening made into the vagina in front of the cervix, the ligatures securing the pedicle carried through it by traction on them, the stump inverted into the vagina, the peritoneum over the inverted cervix closed by stitching the bladder peritoneum to that covering the pedicle, closure of the abdominal wound, and finally placing a special hemostatic clamp on the inverted pedicle in the vagina.

Meinert, independently of Byford, suggested pulling the pedicle into the vagina through Douglas's cul-de-sac, but is not known to have accomplished it.

Polk, of New York, has removed the entire cervix, stitching the vaginal stump to the abdominal wall.

Thus briefly do we get an outline history of the development of the technique of this important operation. From the beginning the struggle was in the direction of accomplishing complete hemostasis of the pedicle without the necessity of invariably fixing it in the abdominal wall. It was soon demonstrated that no pedicle comprised of cervical or uterine tissue could be made bloodless by any amount of ligating with non-elastic ligatures which could not from time to time be tightened as the tissues shrunk. Hence with silk, steel or clamp hemostasis it was necessary to fix the pedicle externally in order that they might be tightened in case of necessity. Elastic ligatures, while they accomplished perfect hemostasis, experience soon demonstrated were not safe ligatures to bury, because they frequently gave rise to suppuration when the strangulated pedicle was dropped. At last it seemed inevitable that external fixation of the pedicle was to be the only safe method of accomplishing abdominal hysterectomy. The displacement of the tissues necessary for abdominal fixation and its distressing sequelae—bladder pressure, painful cicatrix, dragging pain, herniæ, depressed cicatrix, etc., made surgeons slow to accept that means as final, while at the same time there seemed no other alternative. The vaginal fixation of Byford's which came in late in the race, solved many of the difficulties, and if something better had not speedily followed, it would have become the ideal method of pedicle fixation. But when the struggle was at its height the whole problem was suddenly solved by the application of a simple little principle described by Stimson in 1889 and practiced by others and redescribed and emphasized by Baer, of Philadelphia, in 1892. The principle consists in obtaining hemostasis of the uterine stump by ligating its blood supply outside of the uterine tissue before it reaches its substance; or, in other words, by ligating the uterine arteries at either side of the cervix. Eastman had practically accomplished the same thing in his old operation of complete removal of the uterus the same year Stimson announced it, but none of us recognized the principle involved, nor did he announce with sufficient emphasis why he succeeded. So Stimson and Baer get the credit of promulgating and establishing a great but simple principle, and the uterus is removed every day now, partially or wholly, and the pedicle dropped with perfect impunity.

5. *Ligation of Arteries at Side of the Uterine Tissue with Intra-abdominal Stump; Stimson-Baer Method.*—This method is accomplished by ligating the ovarian arteries with or without the broad ligament, severing the broad ligaments after placing hemostatic clamps on the uterine side down to the uterine arteries, ligation of the uterine arteries, severing the uterus at the cervix, cauterizing the cervical canal, trimming the cervix, closing the stump with catgut or silk, burying the pedicle with peritoneum, closing the broad ligaments with a running stitch of catgut, and closing the abdominal wall.

Senn modifies this operation by stripping the tumor of its peritoneum in front and behind for three inches, severing the tumor at its bottom so as to leave the peritoneum like a cuff and then fixing this cuff open to the lower angle of the abdominal wall, draining it with iodoform gauze until all danger of hemorrhage has ceased, when the gauze is removed and the cuff closed by closing the abdominal wound by tying sutures inserted at the time of the operation. Stimson-Baer principle when thoroughly carried out makes Professor Senn's precautions superfluous.

INDICATIONS FOR ABDOMINAL HYSTERECTOMY FOR UTERINE FIBROIDS.

Successful abdominal hysterectomy is the only absolutely sure cure for large fibroids of the uterus. Ergot, electricity, ligation of the blood supply will cure a certain percentage, but hysterectomy removes at once every vestige of the tumor and with it the uterus on which it propagates.

The operation of abdominal hysterectomy, in its present condition of perfection, in the hands of experienced operators should be the operation of selection in all fibroids which can not be removed by vaginal hysterectomy when the patient is in a physical condition which will not jeopardize her immediate recovery from the operation.

Multiple intramural fibroids of every kind which are producing distressing symptoms should be submitted to hysterectomy because there is no absolute cure for them by any other means.

Subperitoneal fibroids when from multiple developments can only be removed by abdominal hysterectomy; no other treatment will reach them.

Interstitial fibroids of large size, of hemorrhagic nature, if the patients are in a fair physical condition, should always be treated by abdominal hysterectomy.

Cystic fibroids can only be cured by hysterectomy. Any form of treatment less radical only aggravates these cases.

Suppurating fibroids imperatively demand hysterectomy.

Fibroids complicated with pelvic suppurations, pyo-alpinx, suppurating ovaries or appendicitis, should be removed at the same time that the pelvis is cleaned out.

Large fibroids complicated with pregnancy where there is the slightest doubt of a successful normal ending of the condition of pregnancy, demand abdominal hysterectomy.

ABDOMINAL HYSTERECTOMY—TECHNIQUE.

Uncomplicated Case.—The writer adopts the Stimson-Baer operation for uncomplicated hysterectomies for any cause. The abdomen is opened with a liberal incision which will allow of easy delivery of the tumor. The lower end of the incision is carried well down to within an inch of the symphysis pubis. If the bladder is unusually high the incision at the lower end need not include the peritoneum. The tumor should next be delivered by lifting it out with the hand or a strong pair of vulsellum forceps fixed in the fundus of the uterus or top of the tumor. It is very necessary that the tumor be delivered at this point in order to continue the work of removal intel-

162

ligently. As soon as the tumor is outside of the abdomen the general peritoneal cavity should be shut off with liberal packs of dry sterilized gauze. If the intestines are inclined to protrude the abdominal incision may be closed above the pelvis with a temporary silk suture. The broad ligaments are next clamped with a strong pair of long jawed hemostatic forceps far enough away from the uterus so that another forceps of the same character may be placed between it and the uterus, and low enough to include all the upper portion of the broad ligament with the ovarian arteries. The broad ligaments on either side are next severed between the forceps to the lower limit of their bite. This frees the uterus well down to the cervix and the region of the uterine arteries. The peritoneum on the anterior surface of the uterus is severed at the utero-vesical fold transversely, the ends of the incision ending at the two provisional forceps placed on the uterine end of the severed broad ligament. The cervix is then stripped of its peritoneum anteriorly, care being exercised to separate the bladder from it, thoroughly. After the uterus is well denuded of its peritoneum below the point marked off by the knife, and the bladder is well separated a gauze sponge of small size may be placed temporarily on the denuded surface. It is well at this point, too, to peel off a small flap of peritoneum from the posterior surface of the lower portion of the body and cervix, beginning an inch above the point at which the stump will be made, and denuding to a point just below it. The uterus is now drawn well to one side, retractors placed on the opposite side and the uterine artery is secured by placing around it a strong silk or antiseptic catgut ligature. The artery is securely tied and the ligature left long. A pair of artery forceps is placed on the tissue secured by the ligature between it and the cervix, and the tissue severed between the forceps and the uterus. The opposite side is treated in the same manner. The uterus is now removed by severing it at its neck. The inci-

sion is begun about an inch above the vaginal attachment anteriorly and posteriorly and carried toward the
uterine canal in such a way as to leave the uterine
stump, a hollow wedge with the apex at the cervical
canal, and the sides of the wedge the anterior and
posterior surfaces of the stump, which when approximated, form flaps which completely shut off the
cervical canal and the cavity of the pedicle from the
abdominal cavity. The uterus can be severed from
the cervix best with a knife. As soon as the flaps

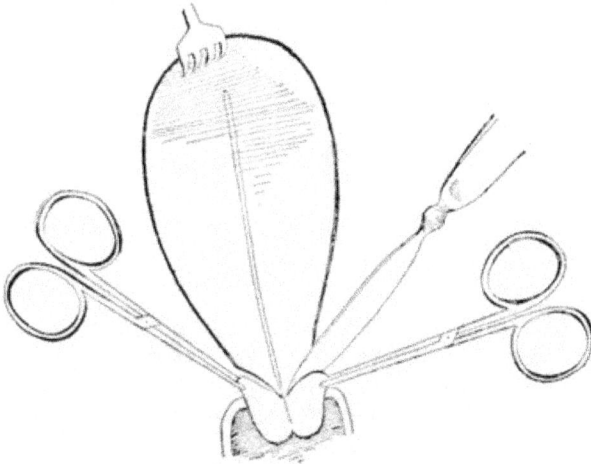

FIGURE 28.

are begun posteriorly and anteriorly the stump should
be steadied and controlled by securing these flaps in
strong lock forceps. (Fig. 28.) As the uterus is severed great care should be exerted not to infect the
abdominal cavity with any septic matter which may
be in the uterine canal, and the cervical canal must
be immediately cauterized or otherwise rendered
sterile.

The stump is now closed by uniting the two flaps with
inversion sutures of antiseptic catgut. The simplest

and most satisfactory method of suturing for this purpose in my opinion is the one employed by Prof. A.
H. Ferguson, Fig. 29. The stitch is an interrupted
one as shown in the drawing, and completely closes
the flaps without penetrating their cut surfaces. Prof.
Ferguson uses the stitch in bowel surgery to take the
place of the Lembert suture. When the pedicle is
closed it is dropped.

At this point the upper portion of the broad liga-

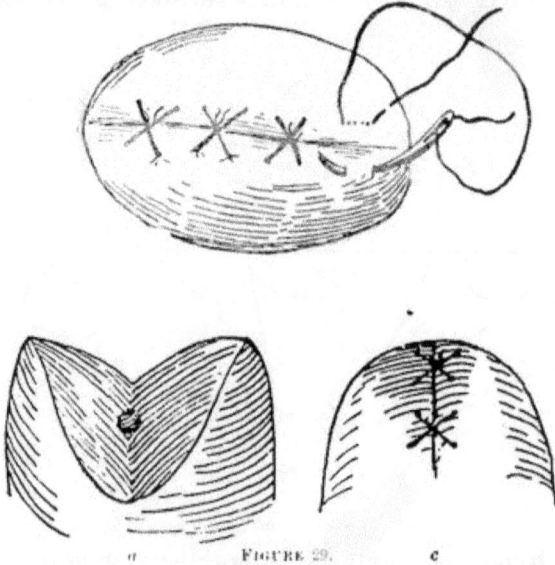

FIGURE 29.

Method of suturing the pedicle in hysterectomy; a, pedicle unclosed;
b, pedicle and method of introducing inversion stitch; c, the completed
pedicle.

ment upon which the provisional hemostatic forceps
were placed must be cared for. It contains the ovarian artery as it passes along the broad ligament parallel to the infundibulo-pelvic ligament which should
be securely tied. Next if the pedicle of the broad ligament is long enough so that it can be easily included in
the ligature left long after ligating the uterine artery

without undue tension, it should be so included, securing the ligature from slipping by taking a turn around the infundibulo-pelvic ligament. When the ligament is in place ready to be tied the forceps should be removed from it and the ligature firmly tied, with a treble knot, and forceps attached to the pedicle outside of the ligature. The opposite side is treated in the same manner.

Everything is now finished about the pedicle except closing the peritoneum over the stump. This is accomplished by stitching together with a running suture of antiseptic catgut the two edges of the peritoneum which was stripped off the anterior and posterior surfaces of the uterus before amputating the uterus. When this is done the pelvic peritoneum is perfectly closed, and as soon as the toilet of the cavity is completed, the abominal wound should be sutured, the dressings applied and the patient put to bed.

Remarks.—The Trendelenburg position may usually be used with advantage immediately after the uterus is amputated in order to expose the bottom of the pelvis. Some operators place their patients in this position from the beginning of the operation. Drainage is not necessary after a normal case.

COMPLICATED CASES.

Unfortunately in the surgery of fibroid tumors uncomplicated cases are not the rule. The most common anomalies are the following: 1, pedunculated tumors; 2, tumors developed into the broad ligament; 3, interstitial tumors involving the cervix; 4, tumors complicated with diseased appendages; 5, suppurating fibroids; 6, tumors complicated with pregnancy: 7, extra-peritoneal fibroids.

1. Pedunculated Tumors.—Tumors of varying sizes with small pedicles are occasionally found growing from some portion of the uterus. If they represent a distinct tumor and the uterus is not involved with separate or other centers of fibroid development, and the appendages are not involved, the operator

should seek to remove the tumor without interfering
with the uterus proper. In order to accomplish the
removal of these pedunculated masses, and secure a
pedicle which may be safely dropped a definite line of
procedure should be followed. If the tumor is only
partially pedunculated so that a portion of its bulk is
buried in the uterus necessitating enucleation, I prefer to remove the uterus, as I consider that the only
absolutely safe procedure under the circumstances. If
the tumor, however, is pedunculated, so that a pedicle
of peritoneum, connective tissue, and the blood vessels
feeding the tumor, without tumor tissue or uterine tissue, can be secured after its removal, I do not hesitate
to ligate and drop the pedicle any more than I would
hesitate to drop the pedicle of an ovarian cyst.

Method: The tumor is delivered. A pair of strong
hemostatic forceps is clamped on the pedicle between
the tumor and uterus, unless the tumor encroaches
upon the pedicle too much to make a clamp effective
after the growth has been enucleated, when instead of
a clamp a provisional rubber ligature should be used.
The tumor should now be cut away. If the pedicle
is long and not involved by the tumor it should be
severed close to the tumor, thus leaving abundance of tissue external to the clamp. If the pedicle
is short and encroached upon by the tumor, the incision should extend around the base of the tumor at a
distance of at least two inches from the provisional
rubber ligature, involving the peritoneal coat and connective tissue capsule of the tumor (the pedicle edges
of the incision being caught on three sides by hemostatic forceps to control the stump and prevent slipping of the ligature), and the tumor is enucleated,
leaving a pedicle of connective tissue, blood vessels
and peritoneum.

The pedicle is treated in both cases alike. If blood
vessels of considerable size are found in the free end
they should be ligated separately close down to the
provisional clamp or ligature. Then a strong antiseptic catgut or silk ligature should transfix the pedi-

cle near its edges, and, after tying the first knot the provisional clamp or ligature should be removed and the ligature securely tied, so as to secure every portion of the pedicle. If care has been observed to eliminate all uterine or tumor tissue from the pedicle it will be as secure now as an ordinary pedicle of an ovarian cyst. The pedicle should be trimmed down to within an inch of the final ligature.

The writer has removed two pedunculated fibroids of large size in this manner, one of ten pounds and another of eight pounds, in which pregnancy existed. the women both going on to term afterward and giving birth to their children without complications. One of these conceived afterward. Another case of this kind in which a tumor of large size was removed, afterward conceived and gave birth without complications to a living child.

2. *Tumor developed into the broad Ligaments.*—It is not a rare complication to find these tumors of the uterus developed in the folds of the broad ligament to such an extent that the ligament is spread out over the growth and its folds tense. Frequently it will be impossible to deliver the uterus until the portion has been enucleated from the broad ligament. The line of procedure here is: First, enucleate the tumor from the broad ligament, and second, deliver the uterus as in uncomplicated cases and complete the toilet in the same manner.

First step: Tie if possible the ovarian artery near the outer edge of the tumor near the pelvic walls. If more convenient a provisional forcep may be used to secure the vessel.

Second step: Split the tense peritoneum which represents the broad ligament, the folds of which have been eradicated by the burying tumor. by drawing a scalpel over it at its most prominent point in a direction from the uterus to the side of the pelvis. Then with the fingers or some blunt instrument the tumor is gradually peeled from its subperitoneal bed, constant traction being exerted on it until the uterus and tumor are delivered.

Third step: Completion of the operation as in a normal case.

3. *Interstitial Tumors involving the Cervix.*—When the fibroid has developed low in the substance of the uterus so as to occupy the cervix, some management is required in order to secure a proper pedicle. Two methods may be pursued: first, complete enucleation of the fibroid tissue from the cervix, and second, complete removal of the cervix.

When it is possible, I prefer to enucleate the tumor from the cervix in order to preserve that portion of the uterus for a pedicle and a key to the abdominal floor.

The first part of the operation is conducted as in a normal case or if any portion of the tumor is subperitoneal as in the last method described. When the region of the large cervix is reached, a blunt instrument should be employed to completely enucleate it from all surrounding tissue, the bladder in front and all lateral tissue in order to insure perfect security of the ureters. This can only be done by keeping the point of the enucleating instrument well against the uterine tissue. If there is difficulty in securing the uterine arteries definitely, because of the necessary distortion of the tissues, two strong Tait pedicle-pins should transfix the cervix, at right angles to each other, their ends being supported by the abdominal walls, and beneath these a provisional rubber ligature should be placed. The tumor is then cut away, down to the pedicle-pins, and the uterine arteries are sought and tied. After insuring hemostasis, the elastic ligature is removed and the fibroid tissue of the cervix carefully peeled out of its capsule. From this point the case is treated as a normal hysterectomy.

Removal of the cervix: Occasionally it may seem best to remove the entire cervix when it is the seat of the fibroid invasion. The same course should be pursued here as when the cervix is to be retained. The uterus may or may not be severed above a provisional elastic ligature, before ligating the uterine arteries. When the cervix has been thoroughly stripped and the vagina rendered aseptic, the vagina should be opened at the anterior or posterior cul-de-sac close to the cervix, and one blade of a curved pair of scissors slipped through into the vagina, and the cervix completely severed from its vaginal attachment by following the circumference of the cervix with the scissors. A guide in the vagina in the form of a staff may be employed in making the first vaginal incision. If there are any small bleeding points on the vaginal edges they should be tied with catgut or twisted with hemostatic forceps. The vagina should be loosely packed with sterilized iodoform gauze from above so as to just reach to the upper end of the severed vagina. The tissues in the bottom of the pelvis naturally fall together. With abundant drain in the vagina I simply allow the upper end of the vagina and the other severed tissues to fall together naturally, contenting myself to close the peritoneum alone with a running antiseptic catgut suture, exactly as when the cervix is allowed to remain.

4. *Tumors complicated with diseased Appendages.*—Diseased appendages are a frequent accompaniment of fibroids of the uterus. When fibroids demanding a hysterectomy are complicated with diseased append-ages the disease of the adnexia should be treated in the ordinary way, and then the hysterectomy should be carried out on the lines best suited to the case.

Cysts of the ovary, without adhesions, scarcely complicates a hysterectomy for fibroids. The pedicle of the tumor is at once ligated with strong silk or clamped with strong forceps and the tumor removed. If it is of considerable size it may first be emptied with a trochar.

Pyosalpinx should be attacked as though no fibroid existed. If there has been bilateral disease of the appendages with extensive peritonitis and numerous adhesions, the adhesions should be carefully separated and the diseased pus tubes and ovaries carefully enucleated and removed. Then the uterus is removed in

the ordinary way. The matter of drainage should be
dealt with here exactly as when no hysterectomy follows, except that it may oftener be more convenient
to drain through the vagina. If enucleation of the
appendages has been such that large peritoneal adhesions have been separated and there is considerable
unavoidable oozing from raw surfaces, some form of
drain is imperative. As the most dependent portion
of the pelvis is Douglas's cul-de-sac, one should select
that point from which to make vaginal drain. Fig. 30
crudely represents an instrument I have devised for
opening the cul-de-sac and guiding my drainage gauze
into the vagina. The lower instrument represents
a staff which is placed in the vagina as a guide, with a

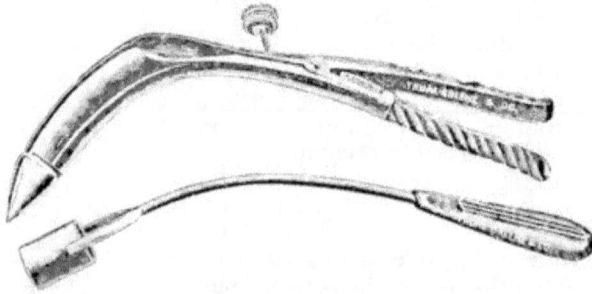

FIGURE 30.

tubular end which will act as a counter pressure for the
pointed dressing guide which penetrates the cul-de-
sac from the pelvic cavity. The upper instrument is
a hollow forceps with pointed blades, which when
they have penetrated into the vagina guided by the
staff are opened and a strip of gauze of any size may be
pushed between them and drawn through from below.
The tubular forceps may also act as a guide for a
rubber drainage tube. Those who have attempted to
place drainage tubes or gauze without a proper guide
will appreciate the advantage of this instrument.

So, after the uterus is removed a roll of sterilized
strip gauze about the size of the index finger should

be drawn through the cul-de-sac into the vagina, the
vagina loosely packed with gauze below and a packing
left in the lower part of the pelvis sufficient to take
care of any oozing from the peritoneal surfaces. Fig. 31.
The toilet of the peritoneum is completed as in ordinary cases and the abdominal wound closed. The
dressing used as a drain is removed as soon as it no
longer soils the dry dressings which are placed in contact with it at the vaginal outlet, usually from twenty-
four to forty-eight hours.

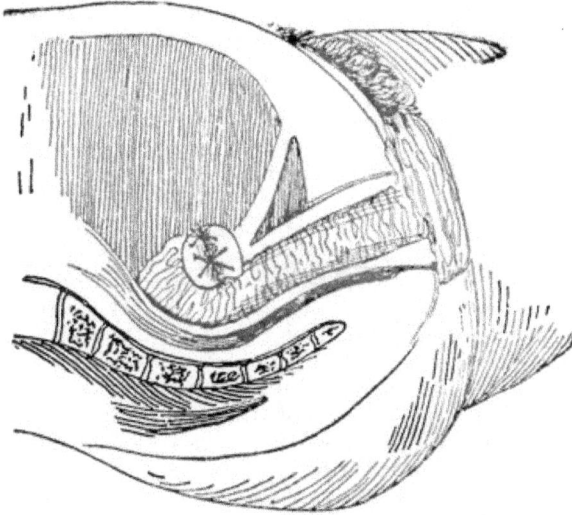

FIGURE 31.

5. *Suppurating Fibroids*—Infected fibroids in
which there is extensive interstitial suppuration are
extremely rare. I have not seen more than two such
cases in my experience. One of these I removed.
The tumor had been infected more than a year before
I operated on it. Several abscesses formed at intervals in the interior of the walls of the large uterus
and then discharged through the cervix. The case

failed to successfully drain after several operative procedures which I attempted through the cervix. Finally I decided to do a complete abdominal hysterectomy, one which would also include the infected cervix. The operation is performed practically as described above for the complete removal of the uterus including the cervix. Extreme care must be maintained to protect the abdominal contents and peritoneum from the infected contents of the uterus. If the vaginal track is employed for drainage, it should be thoroughly cleansed first by a competent assistant who is not allowed to further participate in the operation. All drain gauze should be drawn from above downward, never the reverse.

6. *Tumors complicated with Pregnancy.*— Fibroid tumors of large size complicated with pregnancy demands the sacrificing of the product of conception and the removal of the uterus.

Symptoms: The symptoms of pregnancy are usually all present in an exaggerated form. Menstruation which has heretofore been excessive and frequent on account of the fibroid will cease abruptly. The tumor will begin to grow rapidly. Pressure symptoms are much exaggerated. The bowels and bladder will become crowded and sacralgia and dysuria will result. In a word, all the ordinary symptoms of growing fibroids of the uterus minus hemorrhage, and all the classical symptoms of pregnancy, will become magnified to a painful degree.

Pedunculated fibroids of the subperitoneal variety, with small thin pedicles complicating pregnancy may be removed in the manner described in this lecture under the head of pedunculated fibroids, without disturbing the contents of the uterus. If the tumor involves the uterine walls to any marked degree and the tumor is so large that it will prevent full development of the fetus or its development to the point of viability, or the tumor's position is such that it will interfere with the pregnancy taking its proper course, the entire uterus should be removed with the tumor.

if it is considered necessary to remove the tumor at
once. If pregnancy is known to exist before an operation is determined on for the removal of the tumor, as
a rule it would be safer to empty the uterus as an
early preliminary measure, if it is feasible, reserving
the operation on the tumor for a time after convalescence from the abortion is accomplished.

Operation: However, if it actually becomes necessary to remove a fibroid uterus complicated with pregnancy, either as a matter of choice, or accident from
mistaken diagnoses, the operation is proceeded with
exactly along the lines of an ordinary abdominal hysterectomy. As a rule under these circumstances, the
broad ligaments are loose, and the uterus freely movable making a hysterectomy comparatively easy. Any
complication should be dealt with as in ordinary
cases.

7. *Extra-peritoneal Fibroids.* — It is not infre-
quent that one will find in multiple fibroids that one
or several of the centers of growth have developed
low in the pelvis and in their increase in size they
have gradually elevated the peritoneum and grown
beneath it until they have become actual extra-peritoneal growths. The degree of such complication vary
much in different cases, from a small nodule growing
beneath the peritoneum from the cervix to a tumor
weighing several pounds elevating the peritoneum in
an irregular manner and distorting all the organs of
the pelvis.

Method of Procedure: These cases are all subject
to removal if they are handled in the proper manner.
They must be enucleated. The peritoneum covering
the abdominal surface of the tumor must be carefully
severed at its point of deflection from the tumor on
to the parietes. The tumor should then be grasped
with strong blunt toothed vulsellum forceps and while
traction is being made to deliver the tumor the fingers should carefully enucleate the growth from its
bed. Great care should be observed in order to enucleate it perfectly and free it absolutely from the ure-

· ters or the rectum walls. By following the enucleation the tumor will finally lead to its pedicle which
will be the uterus. The cavity from which it is enucleated should be packed temporarily with sterilized
gauze sponges in order to check serious oozing.
When the tumor is finally enucleated and removed,
together with the uterus, in the ordinary manner, the
work of making a pelvic floor must be accomplished.
If there is not peritoneum enough left to cover the
floor of the pelvis and a large raw surface is inevitable, this should be drained into the vagina by a roll
of gauze an inch in diameter with a packing in the
pelvis sufficiently large to cover the denuded surface.
Occasionally the cavity from which the tumor is enucleated may be packed with gauze and drained into
the vagina as a subperitoneal pocket and the peritoneum closed over it. As a rule these cases require
drainage.

INDEX.

iv INDEX.

www.ingramcontent.com/pod-product-compliance
Lightning Source LLC
Chambersburg PA
CBHW021804190326
41518CB00007B/438